**Torsten Löbner**

# How to synchronize
# the next generation of IPTV

Explantion of the ETSI standardized version

Bachelor + Master
Publishing

**Löbner, Torsten: How to synchronize the next generation of IPTV. Explantion of the ETSI standardized version, Hamburg, Diplomica Verlag GmbH 2012**
Originaltitel der Abschlussarbeit: Implementing ETSI standardised RTCP-based Interdestination Media Synchronization

ISBN: 978-3-86341-262-3
Druck: Bachelor + Master Publishing, ein Imprint der Diplomica® Verlag GmbH, Hamburg, 2012
Zugl. Hochschule für Telekommunikation Leipzig, Leipzig, Deutschland, MA-Thesis / Master, August 2011

**Bibliografische Information der Deutschen Nationalbibliothek:**
Die Deutsche Nationalbibliothek verzeichnet diese Publikation in der Deutschen Nationalbibliografie; detaillierte bibliografische Daten sind im Internet über http://dnb.d-nb.de abrufbar.

Die digitale Ausgabe (eBook-Ausgabe) dieses Titels trägt die ISBN 978-3-86341-762-8 und kann über den Handel oder den Verlag bezogen werden.

# Abstract

In the last years researches in television systems have become more and more a migration from isolated networks to a combined network of internet, telecommunication, television and other services. One of the main aspects in that research topic is to guarantee, that the user experience will be the same or better then the user expects from a common television system

The purpose of this book is to show a possible protocol implementation for a television system with a packet oriented underlying network. The book will show, that ETSI TS 183 063 [1] Annex W specified protocol extension to Realtime Control Protocol will extend a common known network system to reach the goals for users quality needs.

# Contents

# Contents

# Contents

# Glossary

| Notation | Description |
|----------|-------------|
| 3GPP | 3rd Generation Partnership Project |
| | |
| A | Authentication |
| AAC | Advanced Audio Coding |
| AS | Application Server |
| | |
| BC | Broadcast Session |
| BGCF | Breakout Gateway Control Function |
| | |
| CoD | Content on Demand |
| CRS | Content Recommendation Services |
| CSCF | Call Session Control Functions |
| CSRC | Contributing Source Identifier |
| CSV | Comma-Separated Values |
| | |
| DLRR | Delay Since Last Receiver Report Block |
| DLSR | Delay Since Last Sender Report |
| DVB | Digital Video Broadcast |
| DVB-C | Digital Video Broadcast - Cable |
| DVB-T | Digital Video Broadcast - Terrestrial |
| DVD-S | Digital Video Broadcast - Satellite |
| | |
| ECF | Elementary Control Function |
| EFF | Elementary Forwarding Function |
| ETSI | European Telecommunications Standards Institute |
| | |
| FT | Feedback Target |

| Notation | Description |
|---|---|
| GSL | GNU Scientific Library |
| GUI | Graphical User Interface |
| | |
| HD | High Definition |
| HSS | Home Subscriber Server |
| HTTP | Hypertext Transfer Protocol |
| | |
| I-BGF | Interconnect Border Gateway Function |
| I-CSCF | Interrogation - Call Session Control Functions |
| IDMS-RB | Inter Destination Media Synchronization Report Block |
| IMS | Internet Protocol Multimedia Subsystem |
| IMS MGW | Internet Protocol Multimedia Subsystem Media Gateway |
| IP | Internet Protocol |
| IPTV | Internet Protocol Television |
| | |
| LSR | Last Sender Report |
| | |
| MCF | Media Control Functions |
| MDF | Media Delivery Function |
| MF | Media Function |
| MGCF | Media Gateway Control Function |
| MPEG | Moving Picture Experts Group |
| MPEG-TS | Moving Picture Experts Group - Transport Stream |
| MRF | Media Resource Function |
| MRFC | Media Resource Function Controller |
| MRFP | Media Resource Function Processor |
| MS | Media Sender |
| MSAS | Media Synchronization Application Server |
| MSCI | Media Stream Correlation Identifier |

| Notation | Description |
|---|---|
| NGN | Next Generation Networks |
| NTP | Network Time Protocol |
| NTP_TS | Network Time Protocol Time Stamp |
| | |
| OWD | One-Way Delay |
| | |
| P-CSCF | Proxy - Call Session Control Functions |
| PC | Personal Computer |
| PRACK | Provisional Response Acknowledgment |
| PT | Payload Type |
| PTP | Precision Time Protocol |
| PVR | Personal Video Recorder |
| | |
| QoE | Quality of Experience |
| QoS | Quality of Service |
| | |
| RR | Receiver Report |
| RRT | Receiver Reference Time Report Block |
| RSI | Receiver Summary Information |
| RTC | Real Time Clock |
| RTP | Real-Time Transport Protocol |
| RTP_TS | RTP Timestamp |
| RTSP | Real Time Streaming Protocol |
| RTT | Round Trip Time |
| | |
| S-CSCF | Serving - Call Session Control Functions |
| SC | Synchronization Client |
| SC' | Transcoder |
| SCF | Service Control Function |
| SD | Standard Definition |
| SDES | Session Description |
| SDF | Service Discovery Function |
| SDH | Synchronous Digital Hierarchy |

| Notation | Description |
|----------|-------------|
| SDP | Session Description Protocol |
| SIP | Session Initiation Protocol |
| SLF | Service Locator Function |
| SPST | Synchronization Packet Sender Type |
| SR | Sender Report |
| SS#7 | Signaling System No. 7 |
| SSF | Service Selection Function |
| SSRC | Synchronization Source Identifier |
| | |
| TAI | Targeted Advertisement Injection |
| TCS | Time Correlation Stack |
| TISPAN | Telecommunications and Internet Converged Services and Protocols for Advanced Networks |
| TNO | Nederlandse organisatie voor toegepast natuurwetenschappelijk onderzoek |
| | |
| UA | User Agent |
| UE | User Equipment |
| UPSF | User Profile Server Function |
| | |
| VLC | Video Lan Client |
| VLM | Video Lan Media Server |
| VoIP | Voice over Internet Protocol |
| | |
| XML | Extensible Markup Language |

# Nomenclature

| | |
|---|---|
| $i$ | number of the client |
| $n$ | amount of clients |
| $RTPTS_{11i}$ | RTP timestamp of the first packet used for approximation |
| $RTPTS_{12i}$ | RTP timestamp of the second packet used for approximation |
| $RTPTS_{1i}$ | RTP timestamp that belongs to the time that is approximated |
| $RTT_{1i}$ | $RTT_i$ of first packet used for approximation |
| $RTT_{2i}$ | $RTT_i$ of second packet used for approximation |
| $t'_{11i}$ | $t'_{1i}$ of first packet used for approximation |
| $t'_{12i}$ | $t'_{1i}$ of second packet used for approximation |
| $t_0$ | time of send the RTP-packet by media server |
| $t_2$ | time of send $SR_1$ in multicast mode to the clients |
| $t_6$ | time of send $SR'_1$ in unicast mode to MSAS |
| $t_7$ | time of reception of $SR'_1$ by MSAS |
| $t_{1i}$ | time of reception the RTP-packet by $Client_i$ |
| $t_{3i}$ | time of reception of $SR_1$ by $Client_i$ |
| $t_{4i}$ | time of send $RR_1$ by $Client_i$ |
| $t_{5i}$ | time of reception of $RR_1$ by MSAS sent by $Client_i$ |

# List of Figures

# List of Figures

# List of Tables

# 1 Introduction

## 1.1 Preface

This book represents the results of my research in synchronization of television during my graduation project. I will describe a solution, which is actually standardized and give a solution on how to implement it in this document.

It is a pleasure to thank the people who made this book possible. First of all these are my supervisors Oskar van Deventer and Michael Maruschke, who supported my by reviewing my work and discussion on content. I also would like to thank Ray van Brandenburg and Hans Stokking, who were always open for discussion.

## 1.2 Research Purpose and Related Work

This work was done at TNO Information and communication technology. The part of TNO, where this research was done, has its main research topic in media technologies and content delivery systems. Research is done in cooperation with Dutch and international companies as well, as with international research groups. TNO is also a member in the NGNLab project, which main purpose is Next Generation Networks and topics related to that.

## 1.3 Subject of this Book

The purpose of this book is to create a proof of concept of the synchronization system for IPTV described by ETSI TS 182 027 [2] and ETSI TS 183 063 [1] by using the protocol extension to RTCP from ETSI TS 183 063 Annex W. During planing, implementation and evaluation specifications have to be proofed and requirements, for a sufficient work have to be generated, if the standardized environment is not clear defined on some part of the implementation or not sufficient.

This document should give the reader an overview of the necessary requirements and the way of development of the proof of concept.

# 1.4 Books Outline

This book is divided into seven chapters. The first chapters are the theoretical base, followed by the planing and evaluation of the prototyped IDMS system.

In chapter two an overview of the books background and necessary protocols needed for communication is given. This is completed by a description of the network framework, which will be the platform for the synchronization approach. The extension for television usage of the network described in chapter two is explained in chapter three.

The Software analyzed for the usage in the prototyped implementation is described in chapter four. The necessary modifications and extensions to this software and structure of the applications used to build the environment for the described implementation completes the theoretical part of the book. Chapter five shows these software planing.

Chapter six gives and overview of the measurements for proving, that the created implementation works sufficient. This is completed by the summary in chapter eight.

# 1.5 Document conventions

This book follows the following conventions:

- Important terms are italicized when first used, like *Quality of Service* (QoS).

- All code fragments are highlighted like `function` or in an verbatim environment.

- Names of functions are marked with trailing parentheses. Example: **function()**.

- References are made like chapter 2, or figure 2.19, which indicates the type and level the reference points to.

- Figures and tables are numbered on a per chapter basis.

- References to figures and tables are made without page numbers, except the figure or table in question is more than 3 pages away from the reference.

- Footnotes are numbered consecutively from the beginning to the end of the document.

# 2 Theoretical framework

This chapter will give a summarized overview of the background of the book and the theoretical basics needed for planing and implementation of the prototyping implementation.

## 2.1 Next generation TV service

### 2.1.1 Watching TV together

With the progress made in the communication systems around the world, we are now able talk or chat with people on the other side of the world in a high quality. TV broadcasts with sports, movies or TV shows brought people from the past till now to talk about the seen things. Bringing this together to a solution, where the client gets TV and is able to communicate about the contents with friends is the main goal of research on social TV.

The *Telecommunications and Internet Converged Services and Protocols for Advanced Networks* (TISPAN) group, as a part of the *European Telecommunications Standards Institute* (ETSI) has specified IMS-based *Internet Protocol Television* (IPTV) (see: [2] and [1]) as an addition to the *Internet Protocol Multimedia Subsystem* (IMS). This framework supports services for communication and presence as well as services needed for television. These frameworks are discussed later in section 2.3. First of all some background on how TV systems have evolved till now are given in the next point, which is finalized by a summary on quality descriptions for social TV.

### 2.1.2 Evolution of TV

At the beginning of television broadcasts, the content was delivered by as radio signals. Later these signals were also broadcasted via satellite and cable systems. By using these systems there was no need for the content provider to think about quality differences between the receivers. The hole structure was using analog signals and at the clients was no need for a preprocessing. Users of that TV systems got to know that it is possible to talk about the things happened on TV, almost during the show they saw. If there was the same system[1] used, no delay was recognized by the viewers.

The following generation of TV systems used digital system, which need special preprocessing before sending and preprocessing before presentation. One of these systems is

---

[1]radio signal, satellite system or cable system as broadcast system

*Digital Video Broadcast* (DVB). There are three main systems *Digital Video Broadcast - Terrestrial* (DVB-T), *Digital Video Broadcast - Satellite* (DVD-S), *Digital Video Broadcast - Cable* (DVB-C) used, which have different preprocessing. This fact leads to a higher delay between these systems and a delay between different receiver types of the same system could also have a different playout time.

In the mean time data services reached the main bandwidth in communication. At this point the telecommunication networks of the next generation were planed. One of them is IMS, which is a framework inside of the *Next Generation Networks* (NGN). IMS is a very flexible framework for implementation of services and their administration. This high flexibility makes this framework the perfect choice to give the users of such networks the ability to watch TV from the same network as they got voice, chat internet and other services. It also leads to a high interoperability between content providers at low costs. IMS is described in detail in section 2.3.

The usage of NGN depends on a IP-based network, which runs best with a packed oriented lower layer. Data connections where made as data links in synchronized networks like *Synchronous Digital Hierarchy* (SDH) and have now be moved to a packet oriented network. Such networks have one big disadvantage for television services. Data streams are divided into packets, which are transmitted with different delays. This is a result of the per packet scheduling witch is done in packet oriented networks. For further information on that topic see [3].

TV over the internet is called IPTV, which was first used as a simple broadcast of TV content over the internet. At this point it is possible, that users are not viewing the content at the same time. Talking about the content has become a worse, because the opponent of the talk knows things, that will happen later. To face this problem a network has to be designed, that gives the users the known experience of viewing. This makes it necessary to find values for the quality, that could measured and compared to fixed values. As a result there should be a indication of good or bad quality. Two well known quality parameters to solve this problem are described in the next point.

### 2.1.3 Quality needs for Social TV

The main description for quality in communication networks is described by *Quality of Service* (QoS). This is not only one value it is a complete description of parameters of the service. In the past telecommunication networks used synchronous networks, which are optimized for realtime communication. This networks are not very cost-efficient and scalable. For that reasons modern communication networks are packet oriented, which makes them flexible and effective in the usage of their bandwidth. For the usage of telecommunication services over this packet oriented networks new systems for realising QoS are needed and QoS has become one of the most important factors during network development. Delay between sender and receiver, jitter and bandwidth are not only in telecommunication networks important to solve these problems. For television services these values are also important, with the difference that the delay between sender and receiver is not that important then the delay between the receivers, because mostly multicast systems are used.

*Quality of Experience* (QoE) is related to the in the last section described QoS. QoE is no description of the parameters needed by the service, it is the experience the user has during the usage of the service. This could differ between the content of the same service. In practice mostly QoE is measured and QoS parameters are generated as a result of such a measurement.

For social TV the important value for Quality is the delay in presentation between each receiver. This value should lead to the QoS requirements.

*"While currently telecom operators are aiming at the synchronization level found in telecommunication tests (150ms) our results show that voice chatters only start noticing differences above 2 seconds delays. Most text chatters do not notice synchronization differences between 0 and 4 seconds, however active text chatters notice synchronization differences similar to when using voice chat. As the highest levels of togetherness were also observed with active text chatters and all voice chatters, we recommend synchronization of approximately 1 second (which was not noticeable by this group) for a seamless shared experience. These results put into doubt the 150ms value from telecommunications research as the target synchronization bound required for social video watching applications. A first implication for software designers is that they can concentrate on implementing simpler mechanisms that aim at a synchronization level of 1 second (which was not noticeable by this group) for a seamless shared experience. These results put into doubt the 150ms value from telecommunications research [...]"*[4]

This leads to a maximum delay between each receivers presentation of the same video frame of 1s, that has to be evaluated with the solution described in this book.

## 2.2 Protocols

IMS uses a packet oriented network, based on *Internet Protocol* (IP). The necessary protocols for describing an establishment, modification, termination and the main point of view the synchronization are shown in figure 2.1.

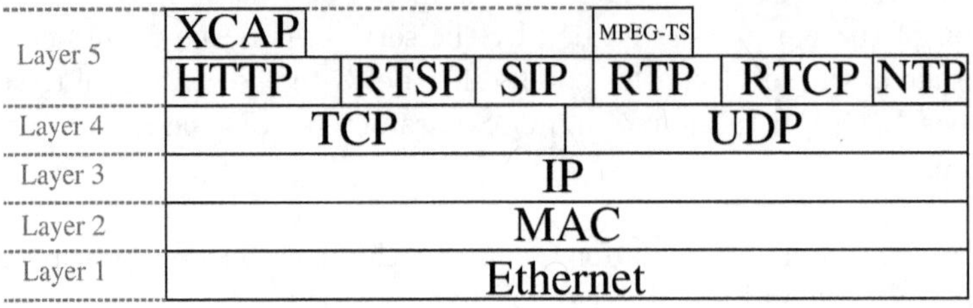

| Layer 5 | XCAP | | | MPEG-TS | | |
|---|---|---|---|---|---|---|
| | HTTP | RTSP | SIP | RTP | RTCP | NTP |
| Layer 4 | TCP | | | UDP | | |
| Layer 3 | IP | | | | | |
| Layer 2 | MAC | | | | | |
| Layer 1 | Ethernet | | | | | |

Figure 2.1: TCP/IP-stack and the protocols for IPTV

These protocols are described in the following subsections.

### 2.2.1 Network Time Protocol

An important step in synchronization is time measurement. To get a correct timestamp all clocks have to be synchronized. One solution for that purpose is *Precision Time Protocol* (PTP) as described in [5]. It describes a solution for clock synchronization with a low error in smaller networks. For synchronizing larger networks *Network Time Protocol* (NTP), as described in [6], is mostly used, because it supports asymmetric delays between server and client. Another advantage to PTP is, that client and server are available as open source software. For the IDMS, described by ETSI TISPAN, it is also mandatory to synchronize all clients clocks using NTP and signaling the source to the application level *Media Synchronization Application Server* (MSAS). How these clocks are synchronized is described in figure 2.2.

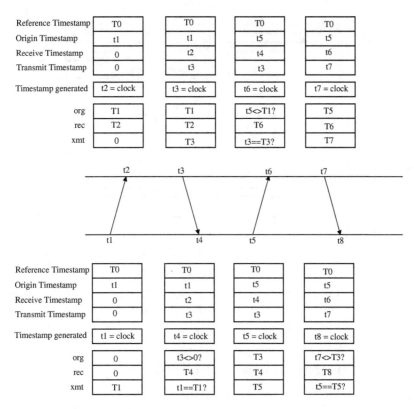

Figure 2.2: Typical clock synchronization using NTP[6]

First the clients sends a packet with its local time (Reference Timestamp - see figure 2.3) and the time of sending the packet (Origin Timestamp - see figure 2.3). The server adds the time of reception (Receive Timestamp - see figure 2.3) and the time of sending the packet back to the client (Transmission Timestamp - see figure 2.3) to the packet and sends it back to the client. The client is now able to calculate a new local time, which is used to do all the step a second time. With all these times the client is able to calculate the new local time, the clock drift and the average computation time of the server. This algorithm is explained in RFC1305[6]. For the IDMS implementation it only necessary to know, that the usage of NTP improves the clock accuracy but it could not be used for every timestamp, because of the bandwidth usage, which leads to an error made, because of the existing clock drift every *Real Time Clock* (RTC) has.

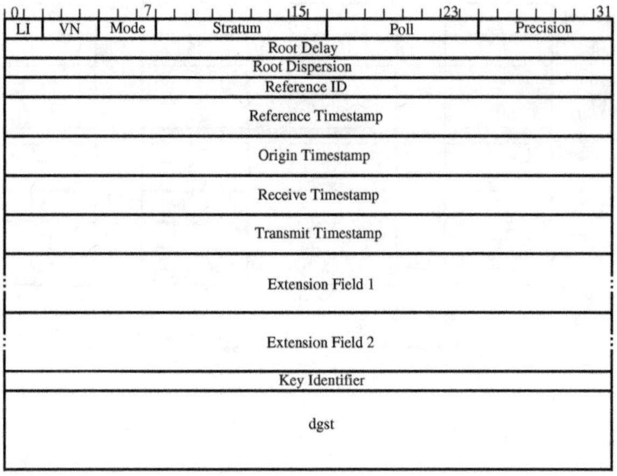

Figure 2.3: NTP frame[6]

Figure 2.3 shows the NTP frame used for clock synchronization. A more detailed description and the description of the necessary local registers (variables) of server and client can be found in RFC1305[6].

## 2.2.2 Session Initiation Protocol

The *Session Initiation Protocol* (SIP) is described by RFC3261 - RFC3265 ([7, 8, 9, 10, 11]). It is mostly used for creation, modification and termination of multimedia sessions. SIP is encapsulated in UPD-datagrams, because of its Three-Way-Handshake, but it is also possible to use it in combination with TCP.

SIP is used for signaling in *Voice over Internet Protocol* (VoIP) based telecommunication systems as well as modified in the IMS. It is the replacement of *Signaling System No. 7* (SS#7) for internet based communication.

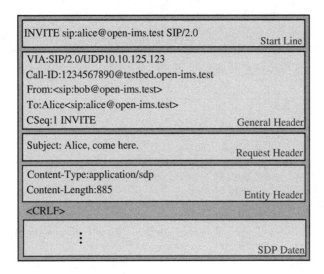

Figure 2.4: Basic message format of the Session Initiation Protocol[7, page 264]

Figure 2.4 shows the INVITE message of SIP. It is like *Hypertext Transfer Protocol* (HTTP) a text based protocol, which makes it flexible in usage. The figure shows the parts of such a message. A detailed description on SIP messages and its contents could be found in [7]. The important part for IDMS, is the content of the session description, which is described in *Session Description Protocol* (SDP) part of the message.

## 2.2.3 Session Description Protocol

For a description of the session the SDP is used. It is described by [12]. This protocol could be used very flexible, because of its ASCII-based architecture it could be modified without modification of the core protocol.

```
v=0
o=- 15115003341359513177 15115003341359513177
    IN IP4 mdf.open-ims.test
s=Unnamed
i=N/A
c=IN IP4 224.0.0.9/255
t=0 0
a=tool:vlc 1.1.8
a=recvonly
a=type:broadcast
a=charset:UTF-8
a=control:rtsp://mdf.open-ims.test:8080/channel1.sdp
m=video 8000 RTP/AVP 33
b=RR:0
a=rtpmap:33 MP2T/90000
a=control:rtsp://mdf.open-ims.test:8080/channel1.sdp/trackID=0
a=rtcp-xr grp-sync,:sync-group=<SyncGroupId>,
```

Figure 2.5: SDP contents for a TV session

Figure 2.5 shows an example description for an IPTV session. The important line of the shown message is `a=rtcp-xr grp-sync,:sync-group=<SyncGroupId>`, which contains the *SyncGroupId* value, which is equal to the *Media Stream Correlation Identifier* (MSCI) used as value for the RTCP communication. A detailed description of the values could be found in [13] and [1].

## 2.2.4  Real-Time Transport Protocol

The *Real-Time Transport Protocol* (RTP) (see: [14]) is used to encapsulate multimedia contents for transmission from the sender to the receiver. For IMS-based IPTV this

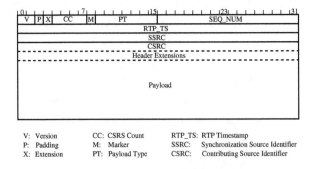

| | | |
|---|---|---|
| V: Version | CC: CSRS Count | RTP_TS: RTP Timestamp |
| P: Padding | M: Marker | SSRC: Synchronization Source Identifier |
| X: Extension | PT: Payload Type | CSRC: Contributing Source Identifier |

Figure 2.6: RTP frame [14, 15]

protocol is also used to transmit the TV contents. Figure 2.6 shows a typical RTP frame. *RTP Timestamp* (RTP_TS) and *Synchronization Source Identifier* (SSRC) are the important values for IDMS in IMS-based IPTV. The RTP_TS is the common base for synchronization between the receivers, because this is the unique identifier for the packet. The packet sender is identified by the SSRC. The *Contributing Source Identifier* (CSRC) field is not used in IPTV, because there is only one media sender. this results in a shortened header without this field.

## 2.2.5 MPEG-TS

*Moving Picture Experts Group* (MPEG) is one of the groups, which standardizes digital formats for video and audio encoding. These formats are used for all DVB systems as well as for IPTV. All these formats could be encapsulated in RTP directly, but this leads to separate data streams for audio and video. For television that could lead to a decrease in performance, because synchronization of audio and video has to be done at the receiver[2]. This problem could be solved by *Moving Picture Experts Group - Transport Stream* (MPEG-TS), which encapsulates audio, video and subtitle streams to a joint stream. The receiver is now able to decapsulate these streams with now need of synchronization[3].

---

[2]The problem of synchronization of separate RTP-streams is described in section 3.2

[3]The synchronization of playout has to be done and is not meant by this.

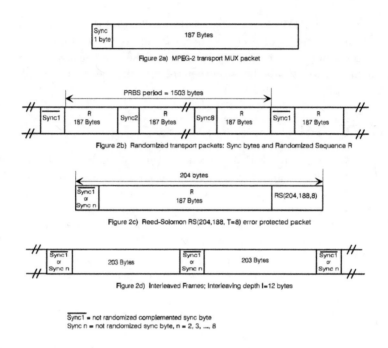

Figure 2.7: Frame structure for mpeg-ts frames [16]

For IMS-based IPTV[4] the MPEG-2 transport MUX packet is used, because error recognition, the advantage of the other frame formats, is done by cheksum in the Ethernet frame (see: [3] for a detailed description).

---

[4]For a detailed description of the packet format see [1]

## 2.2.6 Real-Time Transport Control Protocol

The missing hand-shake system from UDP[17], requests a separate signaling to the sender about the quality of the transmission. RTCP closes this gap, by giving the opportunity to send information of packet sending to the receiver and giving feedback of the reception to the sender. Some of the messages sent by RTCP are described in the following subsections. These are the most interesting ones for synchronization and quality of the session. Important for every RTP / RTCP implementation is the way the packets take in the network. This has to be the same for RTP and RTCP and should not change during the session for best performance, because every route change will lead to inter arrival jitter. This jitter is never zero, because every routing decision takes time, which changes depending on the network traffic (see [18]).

### 2.2.6.1 Sender Report

The Client is able to measure the one-way delay of the transmission, if the sender is configured to send *Sender Report* (SR) messages to the client. These messages are send regularly, but not for every RTP packet.

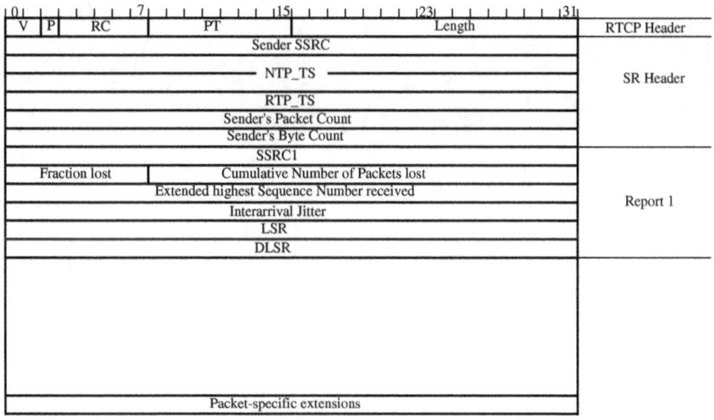

Figure 2.8: RTCP-message: Sender Report[14]

For IPTV only the SR header is interesting, because it contains the combination of RTP_TS and *Network Time Protocol Time Stamp* (NTP_TS), which indicates the exact sending time of the RTP packet it reports on. In a two-way RTP connection, the Sender Report could also be used to send information about the reception to the other participant. Report Blocks are added to the message for each sender to report on. One of the goals for these reports is the measurement of the round trip time. The time stamps *Last Sender Report* (LSR) and *Delay Since Last Sender Report* (DLSR) are used for that calculation. LSR represents the lower 16 Bits of the seconds and the higher 16 Bits of the fractional part of the NTP_TS of the last received SR. The delay between reception of a SR and sending of the current report is represented by the DLSR, that contains the seconds in the higher 16 Bit and the 1/65365 part of a seconds lower 16 Bits. A more detailed description of this delay measurement is described later. A more detailed description about the other contents of this message and there use could be found in RFC3550[14].

### 2.2.6.2 Receiver Report

The *Receiver Report* (RR) contains a report block as described in section 2.2.6.1.

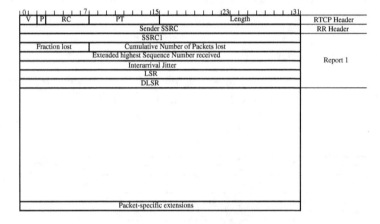

Figure 2.9: RTCP-message: Receiver Report[14]

This message is used in a one-way RTP connection like IPTV, because its one purpose is to report about reception of RTP and RTCP messages.

### 2.2.6.3 Receiver Summary Information

In a multicast environment, the sender sends its RTP stream and RTCP messages to a multicast group with an unknown amount of receivers. If the receiver transmits its RR to the same multicast group, all the other receivers will also receive this message. To save bandwidth it is possible to aggregate receiver information of several receivers to only one message. This avoids sending of unneeded RR to receivers. IP multicast is described by RFC3171[19]. The *Receiver Summary Information* (RSI) is the message which should be used to send aggregated receiver information to the sender.

## 2.2 Protocols

| 0 | | | | | | | 7 | | | | | | | | 15 | | | | | | | | 23 | | | | | | | | 31 | |
|---|---|---|---|---|---|---|---|---|---|---|---|---|---|---|---|---|---|---|---|---|---|---|---|---|---|---|---|---|---|---|---|---|---|
| V | P | RC | | | | PT | | | | | | Length | | | | | | | | | | | | RTCP Header |

The supreport could be one of the following:

- Loss Sub-Report Block,

- Jitter Sub-Report Block,

- Round-Trip Time Sub-Report Block,

- Cumulative Loss Sub-Report Block,

- Feedback Target Address Sub-Report Block,

- Collision Sub-Report Block,

- General Statistics Sub-Report Block,

- RTCP Bandwidth Indication Sub-Report Block or

- RTCP Group and Average Packet Size Sub-Report Block

### 2.2.6.4 Session Description

The participants of the session are able to identify to each other by sending a *Session Description* (SDES). The contents of the chunk, shown in figure 2.10 are all possible values. These values are mandatory, this means they do not need to be included all in the chunk. This message is described in detail in RFC3550[14].

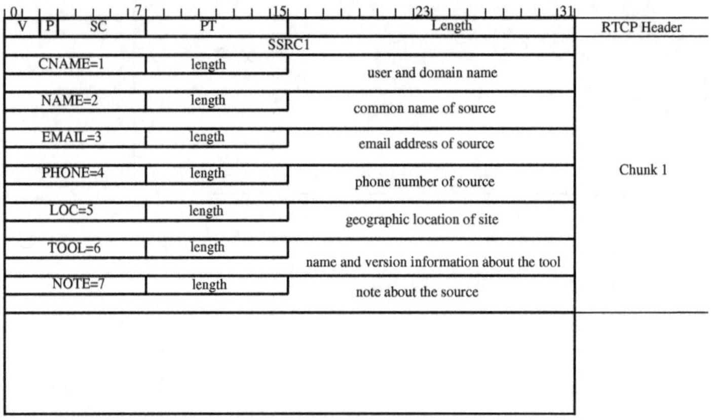

Figure 2.10: RTCP-message: Session Description[14]

### 2.2.6.5 Extended Report

RTCP was designed to transmit status and connection information between sender and receiver of the media content. This behavior does not solve all problems in realtime media systems. Extending these message system is the main goal of RFC3550 [14]. For IDMS some of these messages are important as well. These will be described in the following subsections, finalized by the XR IDMS block described by ETSI TS 183 063 Annex W [1]. The usage of the following report blocks is described later to show their function in common to the used network.

### 2.2.6.5.1 Receiver Reference Time Report Block

One of the functions, that are interesting for IDMS is the delay measurement between non-senders of media described in RFC3611 [20]. RFC3611 describes this as a delay measurement between RTP receivers, but it is also possible to use it in a multicast environment with a feedback target as RTCP receiver for RR. Measurement of the round trip time between feedback target and the receivers is now possible. The *Receiver Refer-*

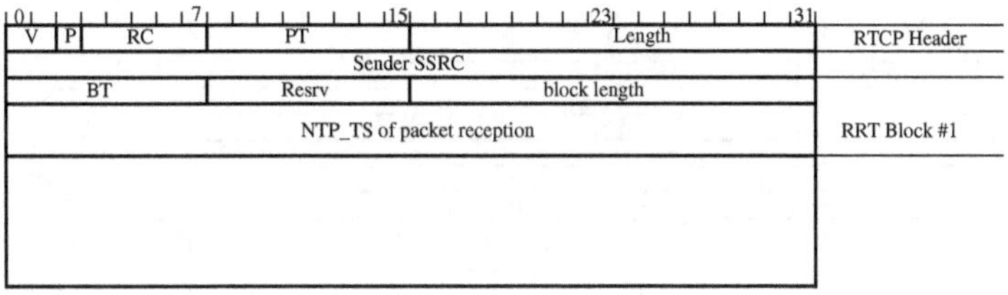

Figure 2.11: RTCP-XR-message: Receiver Reference Time Report Block[20]

*ence Time Report Block* (RRT) is the first XR-block sent for that measurement. It only contains the time of packet sending with a most precise timestamp. This block is shown in figure 2.11 as a full RTCP packet. The other values of that packet are the same as described in section 2.2.6.1 for the SR.

### 2.2.6.5.2 Delay Since Last Receiver Report Block

For the answer of a RRT, as described in a *Delay Since Last Receiver Report Block* (DLRR) is used. Figure 2.12 shows such a Block in a complete RTCP message.

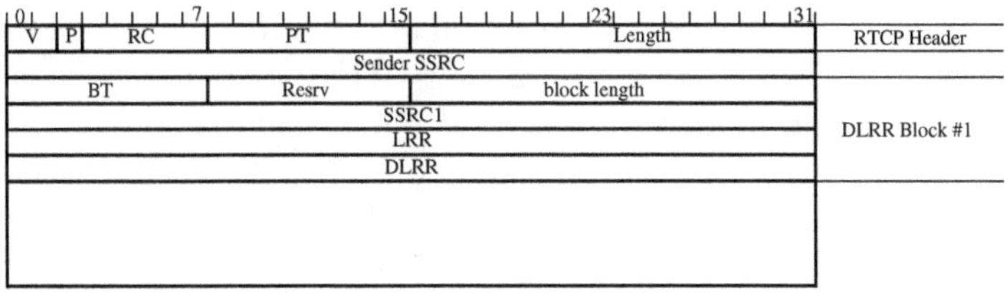

Figure 2.12: RTCP-XR-message: Delay Since Last Receiver Report Block[20]

The SSRC1 values contains the SSRC of the sender of the RRT, to which this DLRR belongs to. The other values are the same as described in section 2.2.6.1 for an RR.

### 2.2.6.5.3 Inter Destination Media Synchronization Report Block

As described in section 2.1.3, the delay in presentation of the media content delivered by RTP is the most important value for the quality of the IPTV system. This leads to the need of sharing the time of presentation between the receivers of the media stream. ETSI TS 183 063 [1] Annex W describes an XR block for sending the time of reception and presentation of the RTP packet.

| 0 · · · · · 7 | · · · · · 15 | · · · · · 23 · · · · · 31 | |
|---|---|---|---|
| V \| P \| RC | PT | Length | RTCP Header |
| Sender SSRC | | | |
| BT \| SPST \| Resrv \| P | block length | | |
| PT | Resrv | | |
| MSCI | | | IDMS Block #1 |
| SSRC of media source | | | |
| NTP_TS of packet reception | | | |
| RTP_TS | | | |
| NTP_TS of packet presentation | | | |

Figure 2.13: RTCP-XR-message: Inter Destination Media Synchronization Report Block[20]

This *Inter Destination Media Synchronization Report Block* (IDMS-RB) as shown in figure 2.13 contains the MSCI of the sync group the client belongs to, the SSRC of the RTP sender, RTP_TS of the packet this reports belongs to and the NTP timestamps of reception and presentation. These are the values that are needed to calculate the delay between several receivers. The *Payload Type* (PT) filed is the same as described by RFC3550 [14] for all RTCP reports. The sender type id identified by the *Synchronization Packet Sender Type* (SPST) field of this message. Table 2.1 shows the allowed SPST for the IDMS reports.

| SPST | Sender type |
|---|---|
| 1 | *Synchronization Client* (SC) |
| 2 | MSAS |
| 3 | *Transcoder* (SC') - receiver |
| 4 | SC' - sender |

Table 2.1: Synchronization Packet Sender Type

### 2.2.6.6 BYE Message

The RTCP session is terminated if one of the parties involved by the session sends a BYE message as described in figure 2.14. This message is not hat important for synchronization, but necessary for the use of RTCP. For that reason it is not described in more detail. The description of the reason for leaving can be found in [14].

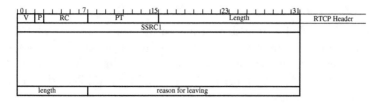

Figure 2.14: RTCP-message: BYE[14]

### 2.2.6.7 Measuring QoS values using RTCP

The monitoring of the QoS values is important for the quality of the IPTV session. RFC3550 [14] describes three values that could be computed by sender and receiver of the RTP session by using RTCP.

The first value is the fraction lost. It is calculated between two SR or RR, that are send by the receiver of the RTP session. This value represents the the number of packets lost divided by the number of packets expected, as shown in equation 2.1.

$$fraction\ lost\ =\ \frac{No.\ packets\ lost}{No\ packets\ expectet} \tag{2.1}$$

[14] This value represents the packet error rate as described by section 2.1.3 between the last sent SR or RR and the actual packet by the RTP receiver.

The cumulative number of packets lost is another value that could be used to compute the quality of the session. This value represents the number of lost packets and is calculated as the number of expected packets subtracted by the number of arrived packets.

A jitter calculation could be done by using the interarrival jitter value, send by the receiver, which represents an estimate of the statistical variance of the RTP data packet interarrival time. Each RTP receiver should calculate this value continuously by using equation 2.3[14]. The delay of each arrived packet should calculated using equation 2.2[14].

$$D(i,j)\ =\ (Rj\ -\ Ri)\ -\ (Sj\ -\ Si)\ =\ (Rj\ -\ Sj)\ -\ (Ri\ -\ Si) \tag{2.2}$$
$$J(i)\ =\ J(i-1)\ +\ (|D(i-1,i)|\ -\ J(i-1))\ /\ 16 \tag{2.3}$$

[14] The *Round Trip Time* (RTT) could be calculated in two ways. First is using LSR, DLSR values and the arrival time of the SR or RR at the media sender by using equation 2.4, taken from [14]. Between non-senders it is possible to use LRR, DLRR and the time of DLRR report block reception by using equation 2.5, taken from [20].

$$RTT_{Sender\ Receiver} = T_{arrival\ SR\ or\ RR\ at\ sender} - DLSR - LSR \qquad (2.4)$$
$$RTT_{between\ non-senders} = T_{arrival\ DLRR} - DLRR - LRR \qquad (2.5)$$

The *One-Way Delay* (OWD) could be calculated by using half of the RTT. Both values are very important for synchronization in IPTV because they indicate changes in the used path between sender and receiver (see: [21]). This indication is only sufficient if the RTT is calculated between sender and receiver, because RTP and RTCP data flow should use the same path in the network.

In a multicast RTP environment the *Feedback Target* (FT) will receive all RR and do all the necessary calculations the sender should do when a RR received and sends the results to the sender by using RSI report block (see RFC5760[22]).

For a QoS measurement application independent passive probes could be used to analyze the RTCP traffic and generate measurement values for a service monitoring. An open source implementation, which describes such a measurement system is described in [23]. This system creates measurement values by probing near to the customer using the values of the RTCP SR. If this system is modified, it could be used to generate statistical information about the performance of the IPTV service, by using SR, RR and IDMS XR messages.

### 2.2.6.8 RTCP architecture for multicast streaming

In most cases IPTV sessions are multicast data streams, because main usage of IPTV is TV broadcast. The usage of IP multicast, has one big disadvantage. For the RTCP session the unicast IP-address is unknown to the receiver at first. This leads to two possible ways of communication. Sending everything to the multicast group is one way for RTCP, but leads to a high bandwidth usage, because RR are send to all participants of that session[5]. Using a separate element for the RTCP traffic back to the *Media Sender* (MS) is a better solution (see. [14]). The FT is used for that purpose.

Figure 2.15 shows this structure. All RTP traffic and RTCP SR are send to a IP multicast group. The receivers of that streams use a unicast transmission to send RR to the FT, which summarizes all reports, by using RSI messages, and sends important information about the clients back to the MS. All QoS measurements could be done by FT, because LSR, DLSR and reception time of the RR are known and the clients report on the other values described in section 2.2.6.7.

In a larger and more flexible network more than one RTP sender and more than one FT

---

[5]See RFC3171 [19] for more details on that topic.

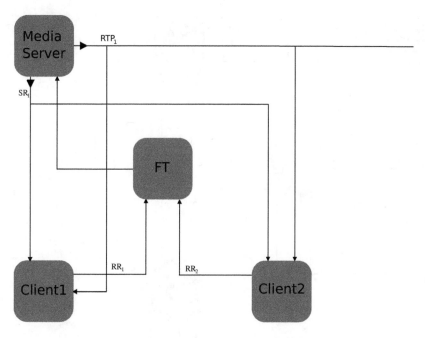

Figure 2.15: Structure for a RTCP multicast environment[22]

are possible for the same content. Figure 2.16 shows a modified version of the structure described by RFC5760 [22]. The new element is the Transcoder, which could be used to provide the same content in another video resolution or audio format. The message flow for RTC is shown in that figure. In a large scale network more than one FT could be used, because every FT gets its own SSRC and sends RSI messages to the corresponding MS.

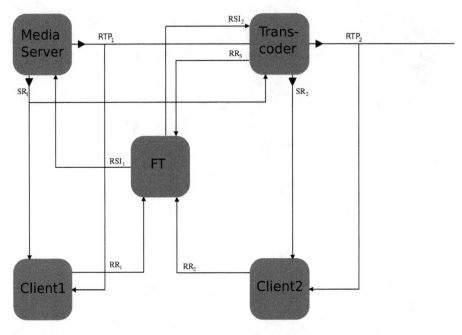

Figure 2.16: Structure for a RTCP multicast environment [22] extend for usage with a Transcoder

# 2.3 Internet Protocol Multimedia Subsystem

The fundament of a scalable, flexible and generally accepted IPTV platform is the standardized network underlying to the needed functions specialized for the TV services. The *3rd Generation Partnership Project* (3GPP) introduced IMS [24], which is a framework for communication, which is scalable and open for new functions. ETSI TISPAN has specified the underlying NGN [25] and actually the IMS of the 3GPP got part of TS 123 228 [26]. Figure 2.17 shows the combined structure from ETSI TS 123 228 [25] and ETSI TS 123 228 [26]

For the usage of IMS in a testbed for IPTV only the *Call Session Control Functions* (CSCF), *Home Subscriber Server* (HSS) or *User Profile Server Function* (UPSF) and the *Application Server* (AS) for the session management are required. A *User Agent* (UA) is used to connect to the core network. The CSCF are divided into *Proxy - Call Session Control Functions* (P-CSCF), *Interrogation - Call Session Control Functions* (I-CSCF) and *Serving - Call Session Control Functions* (S-CSCF).

The IMS function the users client connects to is the P-CSCF, which main functions the

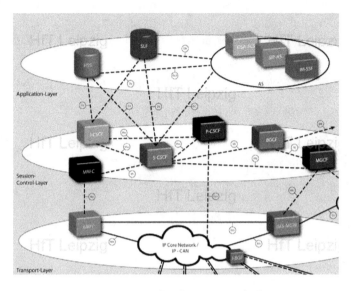

Figure 2.17: IMS overview [27]

locating of the S-CSCF and the HSS and the forwarding of the SIP messages to the S-CSCF, by acting like a proxy. Session control, routing decisions and creation of billing information is done by the S-CSCF. This function is central point in the IMS network, but could be a different one for different services, locations or because of other needs in the network infrastructure. For the connection to other operators the I-CSCF is present in the IMS core network. The provider is able to hide the own network to other operators, which leads to more security, because only the operator knows the network structure and its bottle necks. The mentioned functions done by the CSCF are the important ones for IPTV usage, a more detailed description is given by ETSI TS 123 228 [26].

The main function of the HSS is *Authentication, Authorization and Accounting* (A). Service descriptions and user data are stored in its database. This function is extended to serve the needs of the implemented network. Extensions and network structures are described in 3GPP TS 23.228 / ETSI TS 123 228[28] in detail.

The service ability of the IMS can be increased by using AS, which can act like a UA, a redirect server or a SIP-Proxy. AS do not need to be part of the providers network it is possible to use AS of other providers or of third party distributors, which leads to a very flexible usage of this servers. Connections are redirected to the AS if its functionality is needed. The interfaces to the IMS are described in ETSI TS 123 228 [26].

The *User Equipment* (UE) represents the device the customer uses to communicate with the provider network. The UE consists of the UA needed for communication. All interfaces to the IMS are described in ETSI TS 123 228 [26].

For the interconnection with other networks and operators are the following functions necessary[6]:

- *Internet Protocol Multimedia Subsystem Media Gateway* (IMS MGW)

- *Interconnect Border Gateway Function* (I-BGF)

- *Media Resource Function* (MRF)
    - *Media Resource Function Processor* (MRFP)
    - *Media Resource Function Controller* (MRFC)

- *Breakout Gateway Control Function* (BGCF)

- *Media Gateway Control Function* (MGCF)

- *Service Locator Function* (SLF)

---

[6]See ETSI TS 123 228 [26] for a description of that functions.

## 2.3.1 Basic call

For the communication using IMS, it is necessary to modify the normal SIP procedures for call establishment (see RFC3261[7]) to enable reservation of network resources. This is also very important in an IPTV environment, because the needed network capabilities have to be reserved before the media data are transmitted.

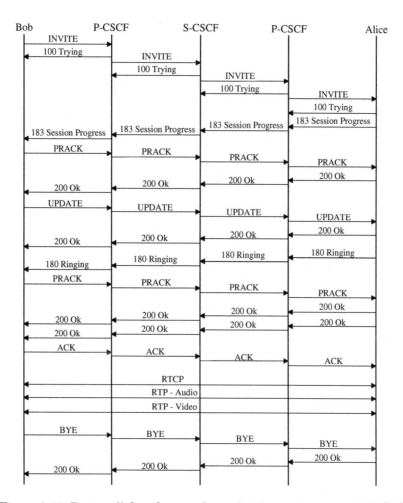

Figure 2.18: Basic call flow for a audio and video session using IMS [27]

In figure 2.18 the basic call flow for establishment, session progress using audio and video streaming and session termination is shown. The interesting parts are the *Provisional Response Acknowledgment* (PRACK) and the UPDATE messages, which are both required to confirm, that the required resources are available. These capabilities were requested with the Session Progress message. By sending the UPDATE message all participants confirm, that the resources are reserved and ready to use. The PRACK message is described in RFC3262[8] and the UPDATE message in RFC3311[29]. The hole call flow is described in [27] or [30]. For the implementation of the synchronization the RTP/RTCP part of that flow is in focus, because this part contains the media control (see section 2.2.6 on page 13).

# 2.4 IMS-based IPTV

As described in section 2.3 it is possible to add new services and additions to the IMS, by adding AS for that purpose. Integration of IPTV is done the same way, with extensions for content delivery.

## 2.4.1 Overview of the Architecture

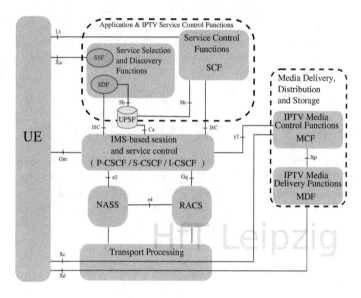

Figure 2.19: Overview of IMS-based IPTV architecture [2]

Figure 2.19 shows this addition to the IMS, to serve the needs for IPTV. The IMS, as described in section 2.3 is represented by the Core IMS, which is the CSCF in most cases during the following descriptions. Information about all services the user is able to use are provided by the *Service Selection Function* (SSF), which responses information about the *Service Discovery Function* (SDF), which belongs to the service, on user request. The SDF is used to provide service information and user based personalized content description to the UE. With the information the client received from the SDF it is able to use the described service.

The *Service Control Function* (SCF) is the AS, which serves user request for the session control of the service. The SCF acts like a UA for the session control, authenticates the user and verifies the users credit-worthiness and payment limits. An other important function is sending relevant information to the *Media Function* (MF). For *Content on Demand* (CoD), *Targeted Advertisement Injection* (TAI), *Personal Video Recorder* (PVR) and *Content Recommendation Services* (CRS) other tasks have to be added to the SCF, these are described in ETSI TS 182 027[2], which also contains a more detailed description on how the described Function interact.

The MF is the important part for the IDMS implementation described in this book, because the functions, it contains of, are delivering and controlling the media content of the session. The MF consists of the *Media Control Functions* (MCF) and the *Media Delivery Function* (MDF). The MCF's main functions are control of the connected MDF and status report of the MDF and the running sessions and services the connected MDF are involved to. Media Server, FT and Transcoder, as described in section 2.2.6.8 on page 20, are implemented in the MDF, which makes this function the most interesting one for synchronization. All interaction with the IMS and the IPTV extensions is done by the MCF, which makes it necessary that the MDF send status, session and QoE information to the MCF. ETSI TS 182 027[2] describes for both functions extensional requirements needed for serving all need in an IMS based IPTV.

## 2.4.2 Watching TV using IMS-based IPTV

In section 2.3.1 on page 25 a standard call flow for establishing and termination of a audio and video session between two UE is described. For IPTV this call flow has to be modified. The following steps show, how a UE can establish, modify and terminate a IPTV *Broadcast Session* (BC) (normal use of TV). All actions in dashed boxes indicate sub flows, which are described in the next points.

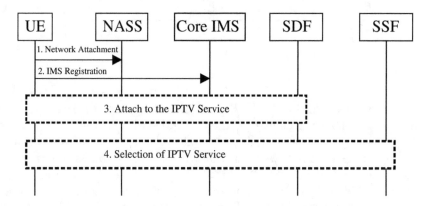

Figure 2.20: Start-up of User Equipment[27][2]

Figure 2.20 shows the start-up of the UE. This step is done the same way as for every IMS registration. The important part is the selection of the IPTV service, which has to be done before connecting to the SCF, which controls the desired broadcast session.

After registration is done and the service description was received by the clients from the

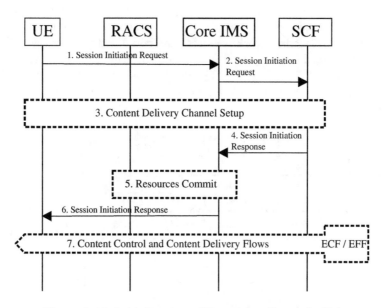

Figure 2.21: Initialization of Broadcast Session[27][2]

SSF, the client is allowed to join a IPTV session, which is shown in figure 2.21. The SCF replaces the called parties UA in the session control, that is why the session initiation Request is forwarded to the SCF by the P-CSCF. The SCF responses, after the content delivery channel is setted, with the Session initiation response. Last shown step are the Content Control (RTCP) and Content Delivery (RTP) flows, which are sent by the *Elementary Control Function* (ECF) and forwarded by *Elementary Forwarding Function* (EFF) of the MF.

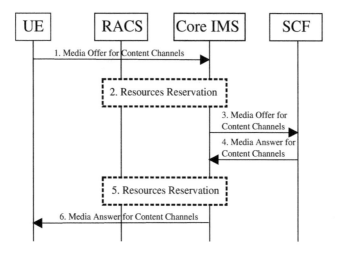

Figure 2.22: Establishing of the delivery channel[27][2]

The previously mentioned delivery channel setup is shown in figure 2.22. This flow is used to reserved the needed network resources and commit this reservation. The client is now able to receive media data (last step in the shown figure). The delivery channel is described by section 2.2.6 on page 13. This is the most important part for the research in this book and will be in focus for the following sections.

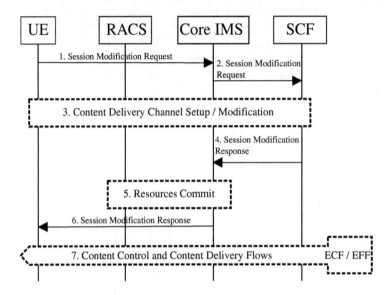

Figure 2.23: Modification of the Broadcast Session[27][2]

During the session it is possible, that the client wants to receive another media stream. This could be done without termination the old and creation of a new session, by modifying the session the way figure 2.23 shows. The content delivery channel is modified the way figure 2.22 describes.

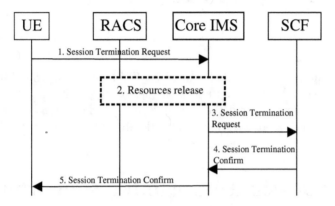

Figure 2.24: Release of the Broadcast Session[27][2]

Finally the session is terminated on user request by using the signaling flow shown in figure 2.24. After that the resources are freed and could be used for other purpose.

The described data flows are taken from ETSI TS 182 027[2] and all signaling is initiated by the UE. A description of other IPTV related services, then BC and a complete set of data flows for all possible actions in a BC session are described in [2] and out of scope of this book. The described steps during a BC session should give an overview of a BC session and show the point in signaling where media data are transmitted.

# 3 Social TV made with IMS-based IPTV

This section will describe a solution for watching TV together by using IMS based IPTV. This will be explained by a description of the necessary modification of the BC session, as described in section 2.4.2 and the signaling flow of the content control and delivery flows. First step is a short description on how IDMS could be solved and how it is solved in IMS based IPTV.

## 3.1 Interdestination Media Synchronization

The common structure for IDMS is shown in figure 3.1. This figure shows the main goal of all algorithms, how to synchronize the playout of the stream at all connected receivers.

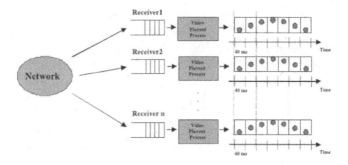

Figure 3.1: Packet flow and presentation overview for group synchronization [31]

In [31] the authors describe a classification of common algorithms for IDMS. This classification contains 4 groups, which are:

- Basic control techniques,

- Preventive control techniques,

- Reactive control techniques and

- Common control techniques.

The implementation of a IDMS algorithm is out of scope of this work, but for testing purpose a simple algorithm to show, that the protocol implementation works. The implementation should contain a pausing algorithm, which belongs to the group of Basic control techniques using receiver control. This type of algorithm is describe by [31] as follows.

> *"Nearly all the solutions use buffering techniques at the receiver side. The reception buffers are used to keep MDUs until their playout instants arrive, according to certain synchronization information, and to smooth out the effects of the network jitter."*[31]

## 3.2 Synchronization of multiple media streams

In combination with IDMS it is also possible to synchronize multiple media streams using RTP and RTCP. If there have to synchronize more then one stream at one receiver and all receiver should be synchronize, only one stream for each client has to be synchronized using IDMS and the others should be synchronized to this. RFC3550[14] gives using the same RTP_TS for the packets of all streams, which belong to each other. This should be also used in a IPTV environment, if the source of all stream is the same sender. In lager IPTV implementations, often these streams are send by different senders, which leads to a complex synchronization of the sources if the same RTP_TS is used.

Using SR is another solution for solving that problem. The synchronization is done after reception of the SR by the sender, because the NTP_TS of of RTP_TS which belongs to each other should be equal[7]. The disadvantage of that algorithm is, that not for all RTP packets SR are sent, which leads to the problem, that one SR from each sender has to be received before the streams are synchronized. This can lead to a long start-up delay, which decreases the QoE.

---

[7]in a real environment they are nearly equal, because of clock asynchronicity.

In an IMS based IPTV environment the RTP_TS of each senders packet, that belongs to the same sending time could be send during session setup. The client is now able to synchronize all streams, because the increase of the RTP_TS is linear for all senders.

# 3.3 Synchronization in IMS-based IPTV

## 3.3.1 Architecture

The following explained structures represent all possible IDMS scenarios in IMS based IPTV standardized by ETSI[2][1]. If all clients receive the same stream and are able to interact with the MSAS the structure shown in figure 3.2 is used.

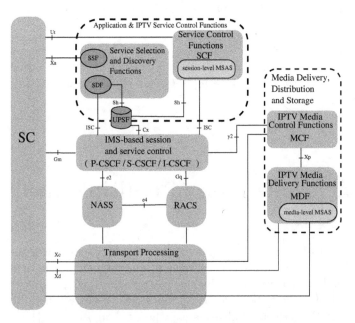

Figure 3.2: Synchronization done between SC and MSAS without Transcoder [27, 2]

In the shown figure the UE is represented by the SC, which interacts with the MSAS for sending and receiving IDMS reports.

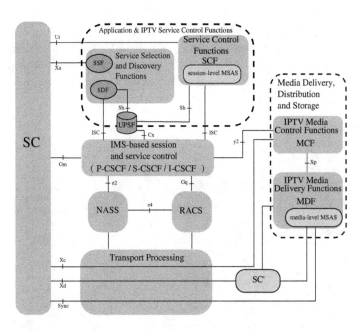

Figure 3.3: Synchronization done between SC and MSAS with Transcoder [27, 2]

If a Transcoder is used, for sending media content exclusively to the user (unicast session) or for sending *Standard Definition* (SD) IPTV stream to a group of users, which is not able to watch the *High Definition* (HD) stream, the architecture shown in figure 3.3 is used. The Transcoder sends information about reception and sending of RTP packets to the MSAS. The RTCP XR reports should contain both IDMS reports on the same RTP frame. This is necessary to determine which incoming RTP_TS belongs to which outgoing timestamp.

In some cases the UE is not able to do synchronization. In this cases the SC could be implemented into the providers network. Figure 3.4 shows this architecture. If the provider chooses this way for synchronization the SC should be as close as possible to the UE in the network architecture. The presentation timestamp could be estimated using the SDES send by the UE, which will work fine for Set Top Boxes (see section 3.4.2 on page 43).

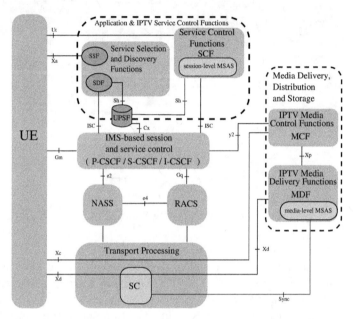

Figure 3.4: Synchronization done on provider side without Transcoder [27, 2]

The fourth possible scenario is shown in figure 3.5. A client is connected to the provider side SC, which receives data from a SC'. In a working environment the provider has to be aware of a mixture of all these scenarios, which makes a synchronization algorithm in the MSAS complex.

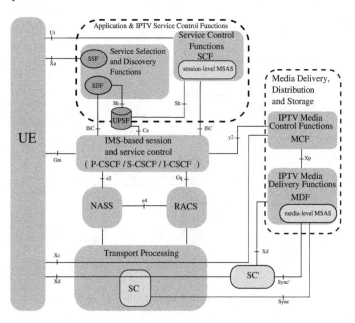

Figure 3.5: Synchronization done on provider side with Transcoder [27, 2]

## 3.3.2 Data flow for synchronization

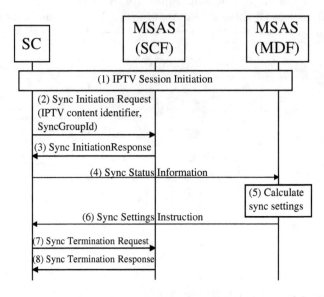

Figure 3.6: Synchroniyation in IMS-based IPTV[2]

The signaling flow for setting up the synchronization and the flows for transmission of the IDMS reports is shown in figure 3.6. Steps 4, 5 and 6 are the most important ones for the implementation of the synchronization, because these steps are necessary to calculate the presentation delay. These signaling is described in the next section, to show the RTCP reports transmission between MSAS and SC.

### 3.3.3 RTCP part of synchronized IMS-based IPTV

For synchronization in IMS-based IPTV an addition to the common used RTCP reports is described in ETSI TS 182 027 [2]. During normal signaling, XR reports for IDMS are sent between MSAS, SC and SC'.

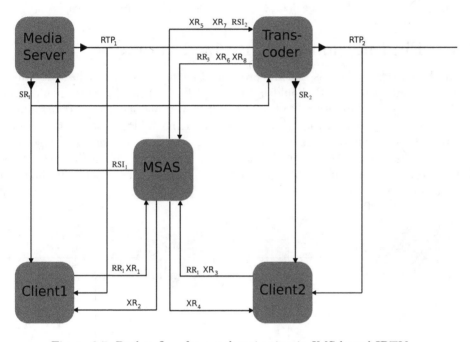

Figure 3.7: Packet flow for synchronization in IMS-based IPTV

The structure, shown in figure 3.7, is an addition to the RTCP multicast architecture, described in section 2.2.6.8. The FT is substituted by the MSAS, which does not only serve its function. In addition the MSAS receives synchronization information and sends synchronization advices back to the SC (shown as XR in the figure).

## 3.4 Getting the clients in sync

For the synchronization it is important to know, how much the difference between each clients clock is or in a simple case to know if the difference in time is to big to synchronize this client. In the following research these two ways, first the indication of a big clock difference and then how to estimate the clock difference, are described. This is followed by some guidelines on how to calculate presentation timestamp and delay.

### 3.4.1 Finding non-synchronized clients

In IMS-based IPTV, there is still the problem, that not every client has the same time base. For that reason NTP should be used on all clients and clients should transmit the NTP-server they are using. But even if this is working fine, there is no specification on how often the clock has to be synchronized to the NTP-server. If this is done to seldom, the clock will system clock will have an offset, which leads to an inaccuracy in the synchronization. In this section a measurement system with existing features of the RTCP protocol is introduced. This could be a solution on the problem by indicating a non-synchronized clock.

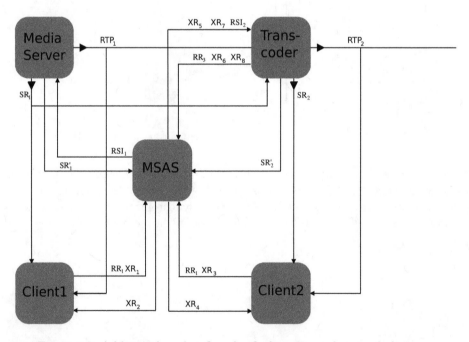

Figure 3.8: Additional packet flow for finding unsynchronized clients

$t_0$          time of send the RTP-packet by media server

| | |
|---|---|
| $t_{1i}$ | time of reception the RTP-packet by $Client_i$ |
| $t_2$ | time of send $SR_1$ in multicast mode to the clients |
| $t_{3i}$ | time of reception of $SR_1$ by $Client_i$ |
| $t_{4i}$ | time of send $RR_1$ by $Client_i$ |
| $t_{5i}$ | time of reception of $RR_1$ by MSAS sent by $Client_i$ |
| $t_6$ | time of send $SR'_1$ in unicast mode to MSAS |
| $t_7$ | time of reception of $SR'_1$ by MSAS |
| $n$ | amount of clients |
| $i$ | number of the client |

These times could not all be measured correctly, because of the difference between the clocks. This difference is from now on expressed by $\Delta$. The difference is a function (as seen below) of the clock drift ($\delta$) and the constant difference between the clock setting ($\Delta_0$). Times influenced by these differences are shown below.

$$
\begin{aligned}
\Delta &\approx \delta \cdot t + \Delta_0 \\
t'_0 &= t_0 + \Delta_{MS\ MSAS} \\
t'_{1i} &= t_{1i} + \Delta_{Client_i\ MSAS} \\
t'_2 &= t_2 + \Delta_{MS\ MSAS} \\
t'_{3i} &= t_{3i} + \Delta_{Client_i\ MSAS} \\
t'_{4i} &= t_{4i} + \Delta_{Client_i\ MSAS} \\
t'_6 &= t_0 + \Delta_{MS\ MSAS}
\end{aligned}
$$

For finding clients, that are not synchronized to a common clock the following relationships are provided. These relationships are depending on the delay (D) between to hosts and there round trip times (RTT).

$$
\begin{aligned}
D_{MS\ MSAS} &= t_7 - t'_6;\ \textit{this delay is equal to all clients} \\
D_{Client_i\ MSAS} &\approx D_{Client_j\ MSAS} \\
RTT_i &\sim D_{MS\ Client_i} \\
\forall_{i=1..n;j=1..n} &\quad i \neq j
\end{aligned}
$$

With that relations it is possible to calculated a common timestamp ($t_x$), which represents the common time between all clients. $T_x[k]$ is a ascending sorted list over all $t_{xi}$, with $T_x[1]$ as the first and $T_x[n]$ as the last element. If $n$ is bigger then ten equation 3.2 should be used for calculating the common time, otherwise there is no need to build the sorted list because equation 3.1 could be used.

$$RTT_i = t_{5i} - t_0' - DLSR$$

$$t_{xi} = t_{1i}' - RTT_i$$

$$t_x = \frac{\sum_{j=1}^{n} t_{xi}}{n} \tag{3.1}$$

$$t_x = \frac{\sum_{j=1}^{n} t_{xi} - T_x[1] - T_x[n]}{n-2} \tag{3.2}$$

$$\Delta_i' = t_{xi} - t_x$$

$$\overline{\Delta_i'} = \frac{\sum_{j=1}^{10} \Delta_i'}{10}$$

$$\delta_i \approx \frac{d\Delta_i'}{dt} \tag{3.3}$$

$$\Delta_{0i} \approx \overline{\Delta_i'} \tag{3.4}$$

An unsynchronized client is indicated by one or more of the following properties:

1. $t_{1i}' < t_0'$, if the MS is synchronized or,

2. $\delta_i > 0$ or,

3. $\Delta_{0i} > 0s$ if $\delta_i \approx 0$

In a real environment the values of $\Delta$ and $\delta$ will not be zero, even if the clocks are synchronized. To prevent wrong indication, these values should be in range around zero. The clients will also not send XR-reports depending on the same RTP-packet. If this happened, equation 3.5 could be used for calculating the time stamp.

$$t_{xi}' \approx \frac{t_{12i}' - t_{11i}'}{RTPTS_{12i} - RTPTS_{11i}} \cdot RTPTS_{1i} - \frac{RTT_{1i} + RTT_{2i}}{2} \tag{3.5}$$

| | |
|---|---|
| $t_{11i}'$: | $t_{1i}'$ of first packet used for approximation |
| $t_{12i}'$: | $t_{1i}'$ of second packet used for approximation |
| $RTPTS_{11i}$: | RTP timestamp of the first packet used for approximation |
| $RTPTS_{12i}$: | RTP timestamp of the second packet used for approximation |
| $RTPTS_{1i}$: | RTP timestamp that belongs to the time that is approximated |
| $RTT_{1i}$: | $RTT_i$ of first packet used for approximation |
| $RTT_{2i}$: | $RTT_i$ of second packet used for approximation |

The last step is the modification of all the equations, to get a good numerical result with less computation. For that reason equation 3.6 is used. That modification also shows, that there is no need to receive SR messages from the Media Server nor from the Transcoder. The common time base is now the RTP time stamp.

$$t'_{xi} \;=\; \frac{t_{12i} \;-\; t_{11i} \;+\; t_{51i} \;-\; t_{52i}}{RTPTS_{12i} \;-\; RTPTS_{11i}} \cdot RTPTS_{1i} \;+\; \frac{DLSR_{2i} \;-\; DLSR_{1i}}{2} \quad (3.6)$$

These values are used to calculate $t_x$ and the values of $\Delta_i$. With these results $\delta_i$ can be calculated using numerical differentiation (as shown in equation 3.7)

$$\delta_i \;\approx\; \frac{\displaystyle\sum_{m=1}^{10} \frac{\Delta'_i[m+1] \;-\; \Delta'_i[m-1]}{2 \cdot T}}{10} \quad (3.7)$$

$\delta_i$ calculated above has the delay between $Client_i$ and the MSAS as error value. This means, a slow receiver will have a higher error than a fast one. To improve this value it is possible to measure the RTT between MSAS an client. This could be done by sending RRT- and DLRR-report blocks. For this measurement the times $t_{0i}$ represent the time of sending RRT-report block to the client and $t_{1i}$ receiving the DLRR-report block from the client. The delay between client and MSAS could now be approximated by equation 3.8

$$\overline{D_{Client_i\ MSAS}} \;\approx\; \frac{\displaystyle\sum_{j=1}^{10} t_{1i}[j] \;-\; t_{0i}[j] \;-\; DLRR_i[j]}{20} \quad (3.8)$$

$$\overline{D_{Client\ MSAS}} \;=\; \frac{\displaystyle\sum_{i=1}^{n} \overline{D_{Client_i\ MSAS}}}{n} \quad (3.9)$$

$$\Delta_{0i} \;\approx\; \overline{\Delta'_i} \;-\; \left[\, \overline{D_{Client_i\ MSAS}} \;-\; \overline{D_{Client\ MSAS}} \,\right] \quad (3.10)$$

$\Delta_{0i}$ could now be used for improving the used IDMS-algorithm, because with the shown calculation $D_{Client_i\ MSAS} \approx D_{Client_j\ MSAS}$ ; $\forall_{i=1..n;j=1..n}\ i \neq j$ is not longer necessary and the result is more accurate to the real difference in time base of the client.

## 3.4.2 Calculation of the Presentation Timestamp

For synchronization of the played out media, the time each RTP-packed is played out has to be known. In an embedded environment, where decoding of audio and video content is done by a hardware implementation this time could be calculated quite easy, because the decoding time is fixed and known, there are no other processes trying to get access to the hardware for decoding and the system is mostly optimized, that means there are no buffer underflows or overflows during playback[8]. This structure is often used in mobile devices and set-top boxes.

The usage of a Software-based decoding leads to the problem of undetermined decoding times. If a hardware based decoder ( for example included into OpenMoko[32], BeagleBoard[33] or PandaBoard[34]) is used for the UE the time of passing the encoded media content to the hardware encoder has to be known to calculate the presentation time, because of the known attributes[9] of the media content the time of decoding could be calculated.

If a SC on provider side is used the session description send by the UE could be used to set an estimated presentation time. with the SDES the client is able to send hardware and software information.

## 3.4.3 Calculation of the presentation delay

The IMS based IPTV allows three possible ways of calculating the pesentation delay, which are divided by the RTP_TS of the sended XR IDMS reports by SC and MSAS. If all clients send RR and XR IDMS reports on RTP_TS, which are reported by SR in multicast, the delay could be calculated without the need of estimation a function. This is neccesary if clients report on different RTP_TS. This scenario is devided into two because the MSAS can send IDMS advices with the same RTP_TS to all client, which makes function estimation on the client side neccesary. Another solution is, that the client sends IDMS advices for each client using the last received RTP_TS. RFC3550[14] and RFC5760[22] do not giv advices on the type of sending RR and XR reports. In a real environment all three possible scenarios are possible depending on the client and MSAS implementation.

---

[8]The mentioned buffer behavior is inside the software, at this point the receiver buffer is not meant by this.

[9]For example bitrate, resolution and used codec.

# 4 Existing Software

The test environment for the IDMS part of IMS-based IPTV is one of the main topics during development of a software implementation. The software needed, to deploy such a testbed is described in this section. One of the main topics on that is, making a decision, which software to use. After choosing the test environment the required software libraries to build MSAS, SC, Transcoder and media sender are described.

## 4.1 IMS Environment (Open IMS)

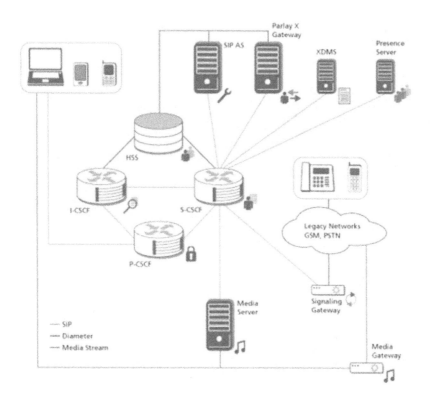

Figure 4.1: overview of the Open IMS structure[35]

Open IMS is one of the most used IMS implementations for academical and research purpose. The shown servers (P-CSCF, I-CSCF, S-CSCF and HSS) are open source and could be installed on a Linux environment. After installation of the Core IMS the example with Alice and Bob, which could be used with the UCT IMS Client is installed. Figure 4.1 shows the Open IMS environment connected possible clients and application servers. A detailed description is available on the homepage of the Fokus Project [35].

## 4.2 Application Server

SCF, SSF and SDF[10] are Application Servers needed to extend the IMS for supporting IPTV services. For the session handling, synchronization information have to known by all participants in that session. To solve that, a part of the MSAS function should be included in the SCF. This part is called application level MSAS. The extension to the Open IMS could be implemented using one of the following Servlet container. An open source application for the Java Servlet and JavaServer Pages technologies is the Apache Tomcat server. It is designed as a HTTP server and supports asymmetric request in the newer versions, which is necessary for SIP communication. The Apache license makes it possible for the programmer to create non-open-source software as extension, which means created Servlets need not to be published as source code. Examples and descriptions on the Apache Tomcat are given by [36].

GlassFish is a Java EE-compatible application server supported by Oracle. This server is available as GPLv2 and CDDL licensed version. This server also supports asymmetric requests.A detailed description of the GlassFish Server and the extensions supported from Oracle are available at [37].

The Mobicents Sip Servlets is an extension for Apache Tomcat and GlassFish, to support SIP signaling using the asymmetric request capability of both server implementations. This extension to the Servlet Container is a certified implementation of the SIP Servlet v1.1 (JSR 289 Spec [38]) specification from Oracle. The usage of the SIP extensions to the Servlet Container is described in *The SIP Servlet Tutorial* [39, 40] and the projects homepage [41].

---

[10]For a more detailed description on that functions see section 2.4

## 4.3 Multimedia players and libraries

The *Video Lan Client* (VLC) is an open source video player and *Video Lan Media Server* (VLM) the corresponding open source media streaming server, which could be used in Content on Demand systems. In newer versions both are implemented as VLC. VLC could be used as library from an application. The full description of VLC is available at [42].

MPlayer is also an open source video player, which supports RTP. This player could be used in embedded environments because there exists special versions for several hardware video decoders. The usage for the SC is very complex because MPlayer is not available as a library. The media player is described in more detail at [43].

All functionality from VLC could be used from within an application using the library of VLC. MPlayer is an open-source software, but only the source code could be reused . The advantage of GStreamer is the flexible structure. Both implementations described before uses a player which could be controlled by sending signals (VLC) or the user has to modify source code of the player to control it from within an application (MPlayer). By using GStreamer, the application designer is able to decide using either an automated decoding and playback or creating a fixed pipeline for encoding and decoding the multimedia data. The elements used for building pipelines in GStreamer are modules which are dynamically loaded during run time. Modules are independent to GStreamer and the interface to GStreamer is fixed, which enables a lot of more flexibility.

| name | description |
|------|-------------|
| filesrc | loads stream from file |
| appsrc | injects stream from application buffer to pipeline |
| decodebin2 | module for automated decoding |
| queue | variable data buffer |
| multiqueue | variable data buffer with multiple inputs and outputs[11] |
| ffmpegcolorspace | converts color space in raw video streams |
| videoscale | modifies resolution in raw video streams |
| capsfilter | blocks streams with wrong capabilities or sets capabilities when unset |
| faac | encodes raw audio streams to AAC |
| aacparse | parses AAC streams to generate capabilities |
| audioconvert | convert resolution and sampling rate of raw audio streams |
| volume | sets volume of raw audio stream |
| alsasink | plays raw audio stream using alsa server |
| ffenc_mpeg2video | encodes raw video to MPEG2 video stream |
| mpegvideoparse | parses MPEG streams to set capabilities |
| mpeg2dec | decodes MPEG2 video streams to raw video streams |
| xvimagesink | plays raw video stream using XVideo output |
| mpegtsmux | multiplexes several audio, video and subtitle streams to MPEG Transport Stream |
| mpegtsdemux | demultiplexes MPEG Transport Stream to audio, video and subtitle streams |
| appsink | passes content from pipeline to the application |

Table 4.1: GStreamer Modules for ETSI IDMS infrastructure

Table 4.1 shows all modules, that are used building the applications for the ETSI IDMS implementation. Their main functions are describe in that table. The documentation of GStreamer and all supported modules could be found on the project homepage [44]. For documentation and easier modification of existing pipelines it is possible for GStreamer to handle with *Extensible Markup Language* (XML) files. The gst-editor [44] is a open source software, that lets the user edit these XML files in a graphical editor.

# 4.4 SIP-Client

## 4.4.1 linphone

The software linphone is an open source SIP UA for every common operating system and smart phone operating systems as well. This UA is not appropriate for the usage in an IMS environment nor an IMS based IPTV systems. The interesting part is the library which is used by linphone and supported by its open source project, because it is a modular library for the RTP and RTCP communication in unicast environments and for the RTP communication in multicast environments. This library is also open source and could be modified for the usage in a multicast RTP environment with unicast feedback. Both, linphone and oRTP are available as ANSI-C source code. The library part is described in detail in the next section. For a test of the modified oRTP library linphone could be used. The reference manual and implementation guidelines could be found at [45].

## 4.4.2 UCT IMS Client

The UCT-IMS-Client is an open source IMS-Client with an extension for IPTV (using VLC). It was developed by the University of Capetown. It's main purpose is to show the communication between IMS-Clients and the OpenIMS from Frauenhofer. For a prototyping implementation of IMS based IPTV this client could be used because all SIP signaling is implemented and the software could be extended. An other advantage to VLC and MPlayer, is the flexible GTK2[? ] environment for the GUI and the modular structure of the C source code module, as parts of this client. The GUI could be modified by using glade [? ], which saves the GUI as an XML file. This makes it possible for the user to modify the look of the software without compiling its source code.

# 4.5 NTP Client

In most cases for synchronizing the system clock superuser[12] right are required to set the clock. If the final application will use this superuser rights, it could become a security risk and in some cases if the user is not allowed to get these rights it could not even started. This leads to a need of a clock synchronization inside the application which sets a time only used inside the application and all its libraries.

For the clock synchronization inside the client application the NTP client implementation created by Larry Doolittle [46] could be used. This stand alone program could be modified to set global variables inside an application and could be run as thread inside that application. This client supports RFC1305[6] for communication and offset calculation.

---

[12]For each operating system this could be called different.

# 4.6 Library for mathematical calculations (GSL)

As described in section 3.4.2 and section 3.4.3, it is necessary to estimate timestamps and calculate differences between these timestamps with a high precision. The *GNU Scientific Library* (GSL) provides a couple of functions for interpolation, approximation and regression. This library is open source, which makes it possible to reuse parts of the source code for the calculation part of the MSAS application. The functions **gsl_fit_linear()** and **gsl_fit_linear_est()** are used to calculate the parameters of the linear function and to estimate the value of the function for a given argument. The coefficients are computed by **gsl_fit_linear()** using linear regression. This function also returns the variance-covariance matrix, which is estimated by using the scatters from the points around the best fit function. These values could finally used to compute the estimation of the functions value by using **gsl_fit_linear_est()**, which also returns the estimated absolute local error. For a more complex mathematical implementation are all functions described by the GSL homepage [47] and the GSL reference manual [48].

# 5 Structure of the Implementation of IDMS for IMS-based IPTV

In the last chapters all theoretical basics were explained, needed to plan the software implementation. This chapter will explain the planing of the IDMS implementation by describing the steps from the modifications of the oRTP library to the Applications needed for profing the concept described in ETSI TS 182 027 [2] and ETSI TS 183 063 [1]. The following description is for the new created functions only. All other functions, if not explained explicitly are described in the reference manual of oRTP (see: [45]).

## 5.1 Library for sending RTP and RTCP data

For the protocol part of the implementation oRTP is used (see section 4.4.1 on page 49). This library is able to use RTP and RTCP for unicast sessions and RTP for multicast session. oRTP has to be modified, for handling XR IDMS report blocks and a RTCP only session, which would be necessary for the MSAS application.

## 5.1.1 Set local address

The following functions are used to set a local address, which is used for RTP receivers (the UE or SC) and RTCP senders (the sending part of the MSAS).

### 5.1.1.1 rtp_synced_session_set_local_addr

For creation of a RTP receiver with unicast or multicast RTCP feedback and SC functionality the function shown in figure 5.1 is used. After determining the IP address family the needed ports are created with given or random port numbers. These created sockets are referenced by the running session in the end of the function, which meant they are accessable for all library functions and if needed from the application.

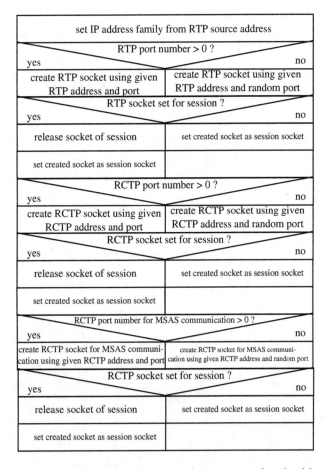

Figure 5.1: Function: rtp_synced_session_set_local_addr

### 5.1.1.2 msas_session_set_local_addr

The function msas_session_set_local_addr sets (figure 5.2) the values for sending for the RTCP only session used by the MSAS application. The function uses the same mechanisms as described for rtp_synced_session_set_local_addr but does not set the sockets for RR and SR communication and the RTP socket.

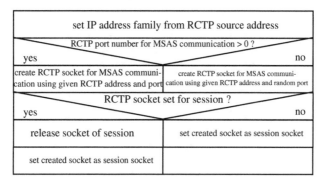

Figure 5.2: Function: msas_session_set_local_addr

## 5.1.2 Set remote address

The following functions are used to create RTCP receiver sockets (for the MSAS) and RTP sender sockets (for the MDF).

### 5.1.2.1 msas_session_set_server_addr

The neccesary socket for the MSAS is created using the function described by figure 5.4. FOr setting the random local adress the function msas_session_set_local_addr could be called from within this function.

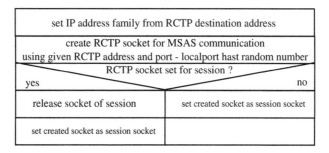

Figure 5.3: Function: msas_session_set_server_addr

### 5.1.2.2 rtp_synced_session_set_remote_addr_full

If the RTP sender should also have MSAS functionality and both are unicast communications, the function described by figure 5.4 could be used. The function creates all necessary remote ports, that the session could be used for a unicast MDF with MSAS functionality.

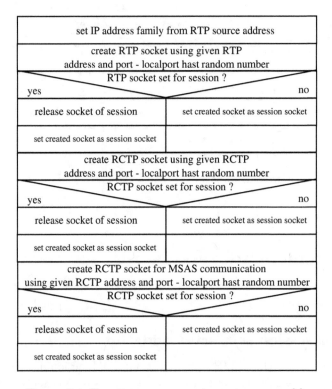

Figure 5.4: Function: msas_session_set_server_addr

## 5.1.3 Create RTCP reports

### 5.1.3.1 rtcp_xr_idms_init

The content of a IDMS reportblock of a XR report is set by using the function decribed by figure 5.5. The contents converted to the report block are taken from a linked list, which contains twice all timestamps. First entry structure of timestamps represents the MSAS contents and the other the SC contents. For the decision, which entry should be used the Sync Type of the active session is used. This function is one of the functions, which should not called from an application, it is used from within the library.

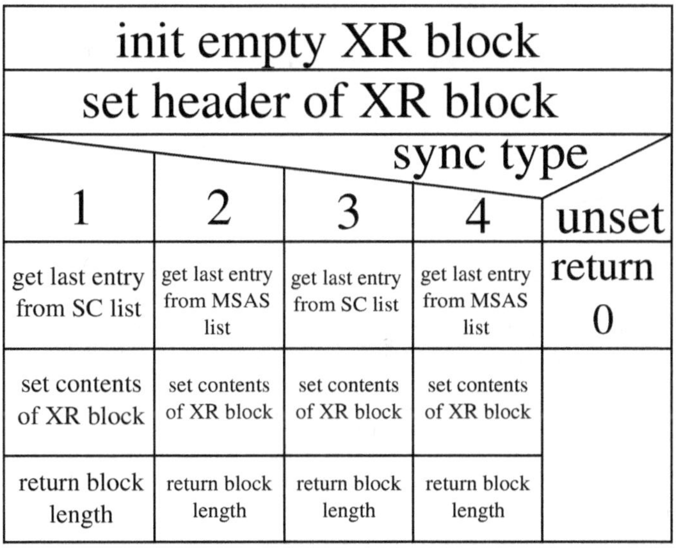

| init empty XR block | | | | |
|---|---|---|---|---|
| set header of XR block | | | | |
| 1 | 2 | 3 | sync type 4 | unset |
| get last entry from SC list | get last entry from MSAS list | get last entry from SC list | get last entry from MSAS list | return 0 |
| set contents of XR block | set contents of XR block | set contents of XR block | set contents of XR block | |
| return block length | return block length | return block length | return block length | |

Figure 5.5: Function: rtcp_xr_idms_init

### 5.1.3.2 make_sr_xr

A compound RTCP message, containing SR, XR and SDES is created using **make_sr_xr()** (see figure 5.6). The SDES is added only, if one or more of the session description is added during initialization of the running session. The SR is created by using **rtcp_sr_init()** and the SDES by using **rtp_session_create_rtcp_sdes_packet()**. This function is one of the functions, which should not called from an application, it is used from within the library.

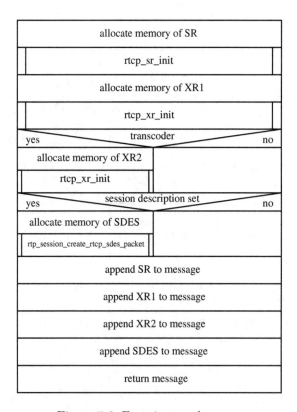

Figure 5.6: Function: make_sr_xr

### 5.1.3.3 make_rr_xr

A compound RTCP message, containing RR, XR and SDES is created using **make_rr_xr()**
(see figure 5.7). The SDES is added only, if one or more of the session description is added
during initialization of the running session. The SR is created by using **rtcp_rr_init()**
and the SDES by using **rtp_session_create_rtcp_sdes_packet()**. This function is one
of the functions, which should not called from an application, it is used from within the
library.

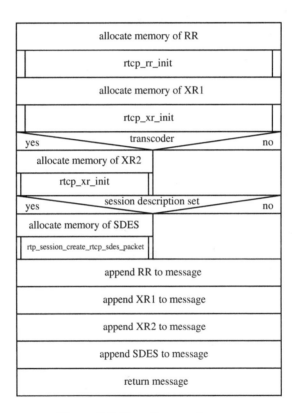

Figure 5.7: Function: make_rr_xr

### 5.1.3.4 make_xr

A compound RTCP message, containing XR and SDES is created using **make_xr()** (see figure 5.8). The SDES is added only, if one or more of the session description is added during initialization of the running session. The SDES is created by using **rtp_session_create_rtcp_sdes_packet()**. This function is one of the functions, which should not called from an application, it is used from within the library.

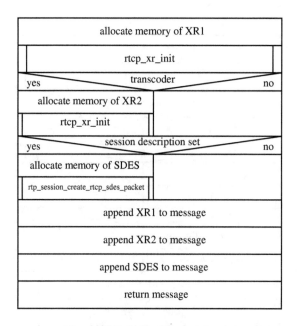

Figure 5.8: Function: make_xr

## 5.1.4 Send RTCP messages

The functions described in this section are used to send all necessary messages for sessions which are synchronized using ETSI IDMS structure.

### 5.1.4.1 msas_session_rtcp_send

The creation of RTCP messages and sending them is done by the function shown in figure 5.9. The functions **session-¿rtcp_sync.tr-¿t_sendto()** and **sendto()** are both system call from the glibc [49] which are used to send the buffer of the socket. This function is called from the library and should not be called from the application.

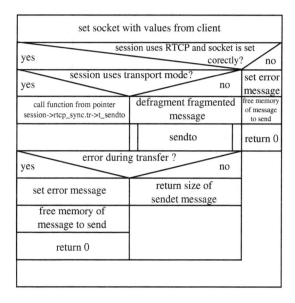

Figure 5.9: Function: msas_session_rtcp_send

### 5.1.4.2 rtp_session_rtcp_sync_send

The creation of RTCP messages and sending them is done by the function shown in figure 5.10. The functions **session-¿rtcp_sync.tr-¿t_sendto()** and **sendto()** are both system call from the glibc [49] which are used to send the buffer of the socket. For sending of the RTP message **rtp_sendmsg()**[13] is used. This function is used in combination of a session created with the function described in section 5.1.2.2. This function is called from the scheduler inside of the library.

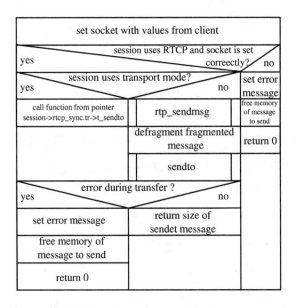

Figure 5.10: Function: rtp_session_rtcp_sync_send

---

[13]This function is described by the reference manual of oRTP[45].

### 5.1.4.3 msas_rtcp_send

RTCP messages are sent to all clients, that have an IDMS advice set by the application, with **msas_rtcp_send()**, shown in figure 5.11. All clients which are known by the MSAS are store in a linked list with its own linked list of timestamps. The function parses the list of known clients, gets its last entry in the time list, creates a RTCP messages and sends it to the client. This function is called from the application at the point all calculations are done.

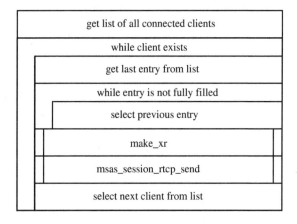

Figure 5.11: Function: msas_rtcp_send

### 5.1.5 Receive RTCP messages (msas_session_rtcp_recv and rtp_session_rtcp_sync_recv)

The functions described by figure 5.12 are used to receive RTCP messages in sessions which are synchronized using ETSI IDMS structure. For a MSAS only session **msas_session_rtcp_recv(** and for a session, where MDF and MSAS are combined **rtp_session_rtcp_sync_recv()** is used. These functions are seperated because **msas_session_rtcp_recv()** is called from the application and **rtp_session_rtcp_sync_recv()** is called by the internal scheduler of the library.

The functions **session-¿rtcp_sync.tr-¿t_recvfrom()** and **recvfrom()** are both system call from the glibc [49] which are used to send the buffer of the socket.

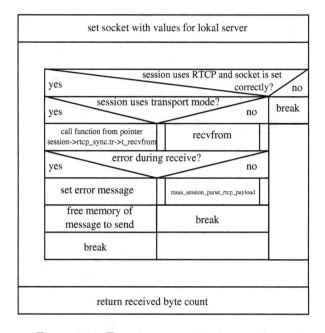

Figure 5.12: Function: msas_session_rtcp_recv

## 5.1.6 Parse content of an XR report block (xr_report_block_parse)

A received XR IDMS report block is converted to entries of the linked list of timestamps by using the function described by figure 5.13. This function is called from inside the library.

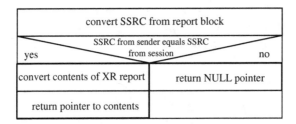

Figure 5.13: Function: xr_report_block_parse

## 5.1.7 Get pointer to the content of an XR IDMS report block (rtcp_XR_sync_get_report_block)

The content of an XR report block is returned, if the size fits to an XR IDMS report block, by the function described in figure 5.14. All sanity checks on the content have to be done by the calling function. This function is called from within the library and should not called from the application.

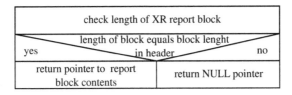

Figure 5.14: Function: rtcp_XR_sync_get_report_block

### 5.1.8 Parse RTCP payload (rtp_session_parse_rtcp_payload and msas_session_parse_rtcp_payload)

For compatibility with the exiting library and the extensions for IDMS, the obsolete function **rtp_session_parse_rtcp_payload()** (figure 5.15) is modified for parsing XR blocks. This function extends the signaling system inside the library. This function is called by the reception functions of the library. The function **msas_session_parse_rtcp_payload()** is exclusively called from the MSAS reception function and sets the values belonging to the MSAS functionality.

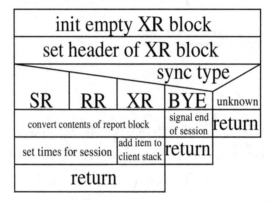

Figure 5.15: Function: rtp_session_parse_rtcp_payload

## 5.1.9 Insert content into linked lists

For the communication between application and library and for sorting contents inside the library linked lists are used. These lists are :

- a list of RTP_TS and NTP timestamps of reception (timestack),

- a list of MSAS and SC based list of RTP, NTP reception and NTP presentation timestamps (TCS)[14],

- a list of all connected clients and

- a list of all known MSCI values as groups of clients.

### 5.1.9.1 push_to_timestack

The list of reception times is filled with **push_to_timestack()** (figure 5.16). If there is an existing item with the same RTP_TS as the one that should be set, the contents are replaced. This function is called during reception function from within the library.

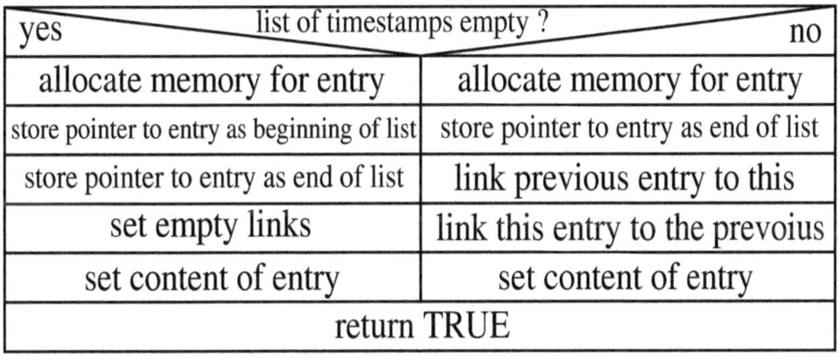

| yes | list of timestamps empty ? | no |
|---|---|---|
| allocate memory for entry | allocate memory for entry | |
| store pointer to entry as beginning of list | store pointer to entry as end of list | |
| store pointer to entry as end of list | link previous entry to this | |
| set empty links | link this entry to the prevoius | |
| set content of entry | set content of entry | |
| return TRUE | | |

Figure 5.16: Function: push_to_timestack

---

[14] *Time Correlation Stack* (TCS)

### 5.1.9.2 insert_ts_to_tcs

If the application has calculated a presentation timestamp for a received RTP_TS, the function **insert_ts_to_tcs()** (figure 5.17)is called by the application. This function copies the contents of the timestack, which belongs to the RTP_TS to the TCS, adds the presentation timestamp and deletes older entries of the timestack.

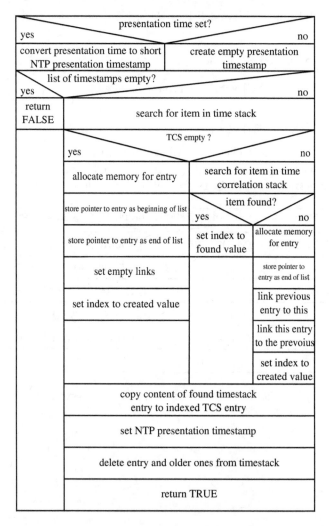

Figure 5.17: Function: insert_ts_to_tcs

### 5.1.9.3 insert_into_tcs

In case a session without RTP reception is active, no timestack entries are created and could not copied into the TCS. In that case the entries could set by the application using **insert_into_tcs()** (figure 5.18).

| yes | tcs empty ? | no |
|---|---|---|
| allocate memory for entry | | allocate memory for entry |
| store pointer to entry as beginning of list | | store pointer to entry as end of list |
| store pointer to entry as end of list | | link previous entry to this |
| set empty links | | link this entry to the prevoius |
| set content of entry | | set content of entry |
| | return TRUE | |

Figure 5.18: Function: insert_into_tcs

### 5.1.9.4 set_tcs_item

If only the SC or the MSAS part of the TCS entry schould be set the function **set_tcs_item()** (figure 5.19) is used. The contents of the TCS is only modified if an entry exists.

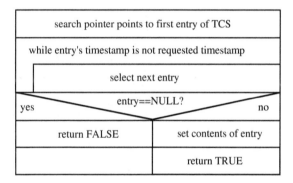

Figure 5.19: Function: set_tcs_item

### 5.1.9.5 msas_session_add_client

This function adds a client to the list of known clients and modifies the list of known MSCI values in an MSAS only session.

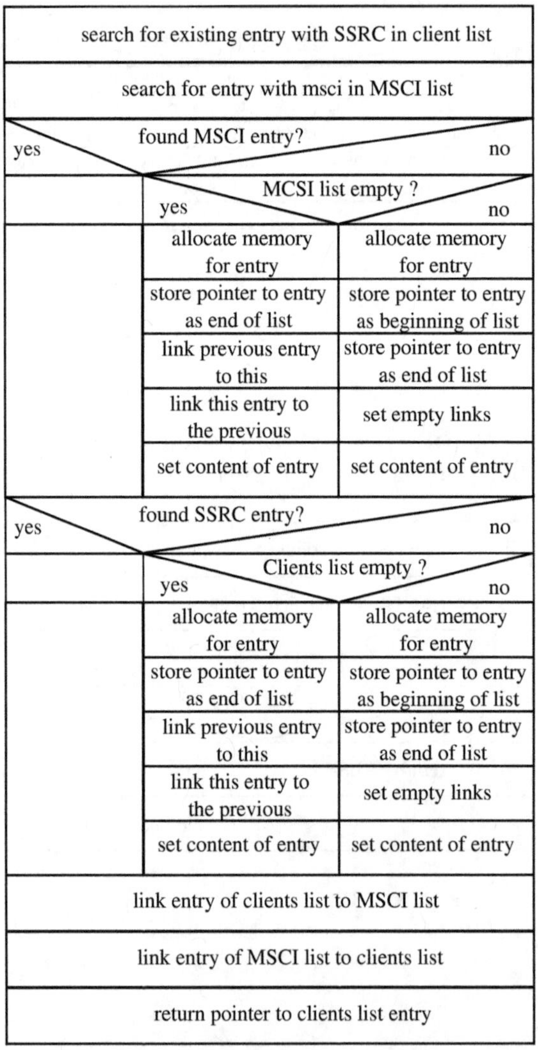

| | | |
|---|---|---|
| search for existing entry with SSRC in client list | | |
| search for entry with msci in MSCI list | | |
| found MSCI entry?    yes / no | | |
| | MCSI list empty ?   yes / no | |
| | allocate memory for entry | allocate memory for entry |
| | store pointer to entry as end of list | store pointer to entry as beginning of list |
| | link previous entry to this | store pointer to entry as end of list |
| | link this entry to the previous | set empty links |
| | set content of entry | set content of entry |
| found SSRC entry?   yes / no | | |
| | Clients list empty ?   yes / no | |
| | allocate memory for entry | allocate memory for entry |
| | store pointer to entry as end of list | store pointer to entry as beginning of list |
| | link previous entry to this | store pointer to entry as end of list |
| | link this entry to the previous | set empty links |
| | set content of entry | set content of entry |
| link entry of clients list to MSCI list | | |
| link entry of MSCI list to clients list | | |
| return pointer to clients list entry | | |

Figure 5.20: Function: msas_session_add_client

### 5.1.9.6 msas_session_insert_tcs_item

For inserting a item to the TCS in a MSAS only session the function **msas_session_insert_tcs_item()** (figure 5.21) is used. This function modifies the function described in section 5.1.9.3 on page 67 to determine the correct client before setting the item.

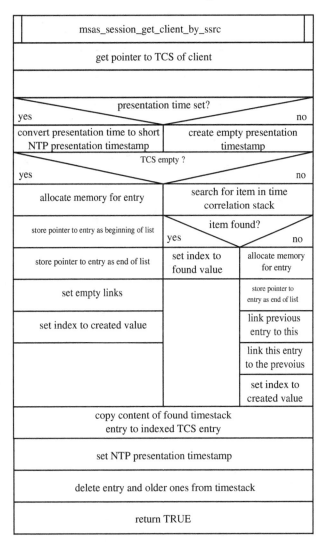

Figure 5.21: Function: insert_into_tcs

### 5.1.9.7 msas_session_insert_tcs_item_msci

The function **msas_session_insert_tcs_item_msci()** (figure 5.22) is used to set the same TCS item for all known clients with the same MSCI value.

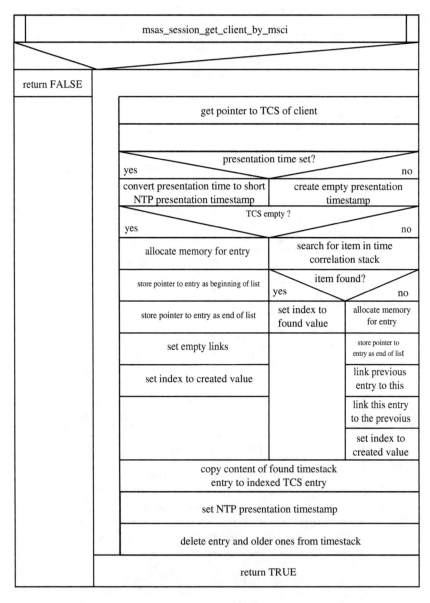

Figure 5.22: Function: msas_session_insert_tcs_item_msci

## 5.1.10 Get contents of linked lists

For getting the items of the linked lists the following functions are used.

### 5.1.10.1 get_item_from_timestack

The item from the timestack can accessed by calling **get_item_from_timestack()** (figure 5.23) for a given RTP_TS.

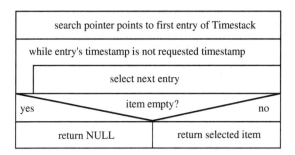

Figure 5.23: Function: get_item_from_timestack

### 5.1.10.2 get_item_from_tcs

The item from the timestack can accessed by calling **get_item_from_tcs()** (figure 5.24) for a given RTP_TS.

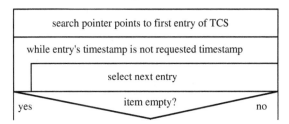

Figure 5.24: Function: msas_session_insert_tcs_item_msci

### 5.1.10.3 get_last_item_from_tcs

The last item of the tcs is returned by calling **get_last_item_from_tcs()** (figure 5.25).

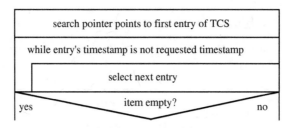

Figure 5.25: Function: get_last_item_from_tcs

### 5.1.10.4 msas_session_get_clients

The pointer to the first or last entry of the list of clients is return by **msas_session_get_clients()** (figure 5.26). This represents the hole list, because all other entry are accessible by using the pointers to the linked elements.

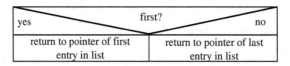

Figure 5.26: Function: msas_session_get_clients

### 5.1.10.5 msas_session_get_client_by_ssrc

The function **msas_session_get_client_by_ssrc()** (figure 5.28) is used to access a client with a specific SSRC.

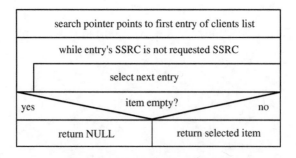

Figure 5.27: Function: msas_session_get_client_by_ssrc

### 5.1.10.6 msas_session_get_client_by_msci

The pointer to the first entry of a list of clients with the same MSCI is return by **msas_session_get_client_by_msci()** (figure 5.26). If no clients are known with the MSCI this function returns an empty pointer.

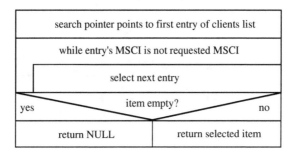

Figure 5.28: Function: msas_session_get_client_by_msci

### 5.1.10.7 msas_session_get_tcs_item

The function **msas_session_get_tcs_item()** (figure 5.29) is a modification of section 5.1.10.2. Before searching the entry the correct client is search and its TCS is selected.

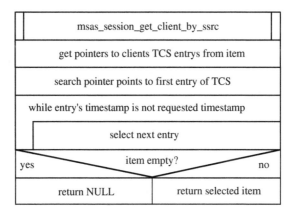

Figure 5.29: Function:msas_session_get_tcs_item

### 5.1.10.8 msas_session_get_last_tcs_item

The function **msas_session_get_last_tcs_item()** (figure 5.30) is a modification of section 5.1.10.3. Before searching the entry the correct client is search and its TCS is selected.

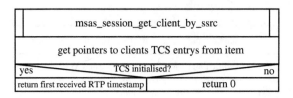

Figure 5.30: Function: msas_session_get_last_tcs_item

## 5.2 Media Delivery Function - RTP-sender part

The MDF is divided into the media delivery and the MSAS, as described in section 3.3 on page 34. In this section the application needed for sending the media stream is described. The application is divided into 3 parts. The MPEG2-TS encoder, which loads media files, transcodes them to MPEG2-TS and passes them to the RTP-sender part of the application. Finalized is the application by the *Graphical User Interface* (GUI) part. These parts are implemented as threads, which are initialized and started by the **main()** function.

### 5.2.1 GStreamer Pipeline

For a flexible usage, the MDF schould support as many as possible media sources. The GStreamer module *decodebin2* serves this need. This module is part of the GStreamer pipeline shown in figure 5.31. The output is passed to the MPEG2 encoder and the *Advanced Audio Coding* (AAC) encoder. Both encoder paths are connected to the *mpegtsmux* module, which muxes the final stream and passes it to the application using the *appsink* module.

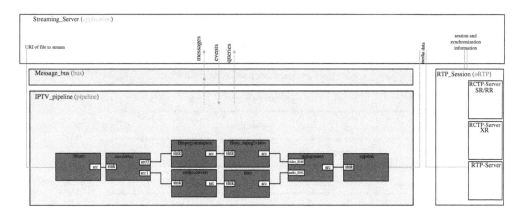

Figure 5.31: GStreamer pipeline for server application

## 5.2.2 Thread for media encoding (encoder_thread_run)

The GStreamer pipeline, described in section 5.2.1, is initialized by the function shown in figure 5.32. After that initialization all modules are connected and configured. Finally the pipeline is started and the thread runs until it is stopped by user request using the GUI.

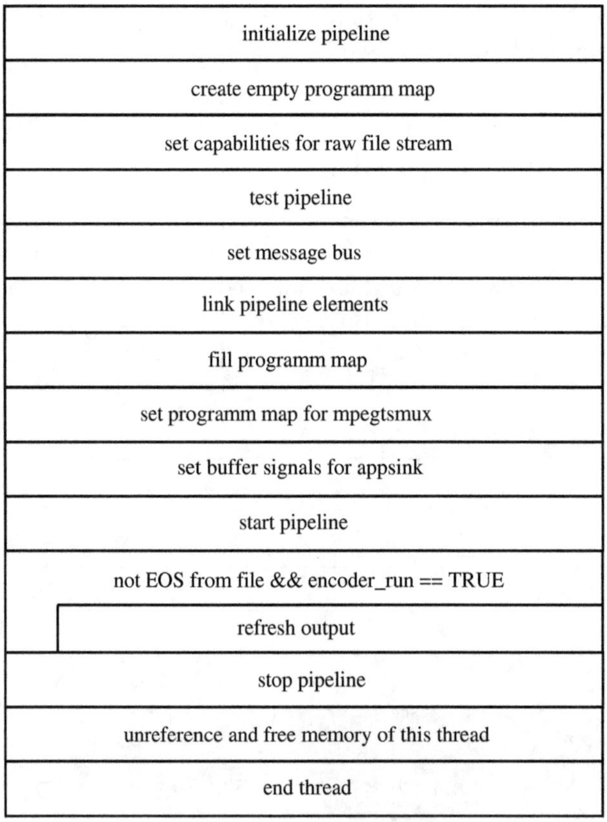

| |
|---|
| initialize pipeline |
| create empty programm map |
| set capabilities for raw file stream |
| test pipeline |
| set message bus |
| link pipeline elements |
| fill programm map |
| set programm map for mpegtsmux |
| set buffer signals for appsink |
| start pipeline |
| not EOS from file && encoder_run == TRUE |
| refresh output |
| stop pipeline |
| unreference and free memory of this thread |
| end thread |

Figure 5.32: Media server - media encoder thread

### 5.2.3 Graphical user interface (gui_run)

The MDF's usage is mainly in a non graphical environment, which makes it necessary to use a console based output for status information. Figure 5.33 shows the thread function for the graphical console output. The output is formatted by using ncurses[15]. By pressing F9 the user is able to exit the application.

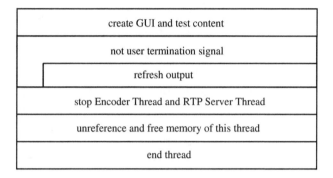

Figure 5.33: Media server - gui thread

---

[15]The usage of this library is explained by [50].

## 5.2.4 RTP-sender (rtp_server_run)

After initialization this function runs in an infinite loop until the users terminates the application or the end of the media file has reached. During that infinite loop the buffer send from the GStreamer pipeline is copied to the send buffer of the oRTP library and send. This is shown in figure 5.34.

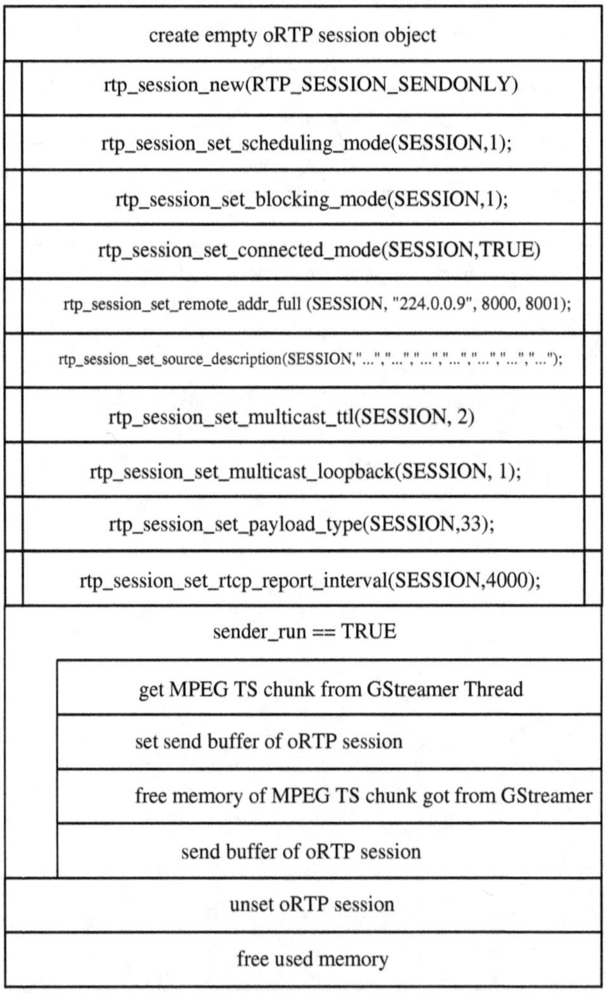

| create empty oRTP session object |
|---|
| rtp_session_new(RTP_SESSION_SENDONLY) |
| rtp_session_set_scheduling_mode(SESSION,1); |
| rtp_session_set_blocking_mode(SESSION,1); |
| rtp_session_set_connected_mode(SESSION,TRUE) |
| rtp_session_set_remote_addr_full (SESSION, "224.0.0.9", 8000, 8001); |
| rtp_session_set_source_description(SESSION,"...","...","...","...","...","...","..."); |
| rtp_session_set_multicast_ttl(SESSION, 2) |
| rtp_session_set_multicast_loopback(SESSION, 1); |
| rtp_session_set_payload_type(SESSION,33); |
| rtp_session_set_rtcp_report_interval(SESSION,4000); |
| sender_run == TRUE |
| get MPEG TS chunk from GStreamer Thread |
| set send buffer of oRTP session |
| free memory of MPEG TS chunk got from GStreamer |
| send buffer of oRTP session |
| unset oRTP session |
| free used memory |

Figure 5.34: Media server - RTP sender thread

## 5.2.5 Primary function of the application (main)

Figure 5.35 shows the main function of the application. This function is started as default by running the application and initializes and runs the threads for the server.

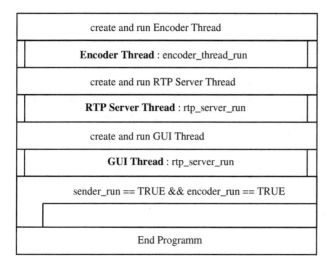

Figure 5.35: Media server - main function

# 5.3 Media Delivery Function - MSAS part

The second part of the MDF is the MSAS, which is described in this section. This server is used as the RTCP Feedback Target and it could be used with more then one RTP senders, which makes it flexible in the usage.

## 5.3.1 Graphical user interface (gui_run)

The MDF's usage is mainly in a non graphical environment, which makes it necessary to use a console based output for status information. Figure 5.36 shows the thread function for the graphical console output. The output is formatted by using ncurses[16]. By pressing F9 the user is able to exit the application.

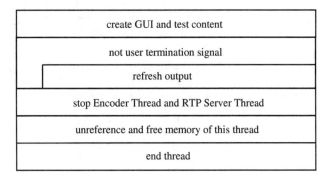

Figure 5.36: Media server - gui thread

---

[16]The usage of this library is explained by [50].

## 5.3.2 RTCP server thread (msas_server_run)

The function described by figure 5.37 is used to create and run the RTCP server. During the runtime the received timestamps of all connected clients with the correct MSCI are used to estimate a linearization function. This functions are used to elect the slowest receiver and send IDMS advices to all connected clients with the same MSCI using the timestamps of the elected client. For each client the advice is send calculated with the last RTP_TS received. The result is, that the client is able to calculate the delay using the difference of the last send and the last received IDMS message.

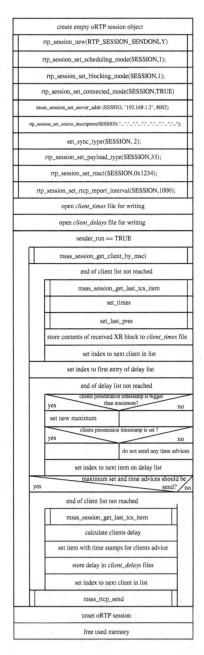

Figure 5.37: RTCP server thread of the MSAS

### 5.3.3 Primary function of the application (main)

Figure 5.38 shows the main function of the MSAS. This function is started as default by running the application and initializes and runs the threads for the RTCP server and the GUI.

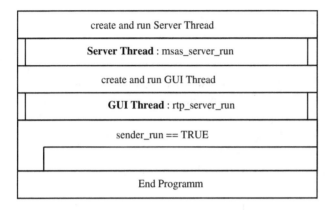

Figure 5.38: Main function of the MSAS

# 5.4 SC application on user side

For the client Side of the IDMS implementation the UCT IMS Client is used. The following described functions are the extension to the existing client.

## 5.4.1 GStreamer Pipeline

The client should be able to decode and present media streams according to the ETSI TS 182 027 [2] specification. Figure 5.39 shows a GStreamer pipeline for decoding MPEG2-TS payload passed to the pipeline using the *appsrc* module. For the measuring of the presentation time the buffer probe of the *xvimagesink* is used. On first received data on that sink buffer the time is measured and stored to determine the delay between reception and presentation of the first RTP packet. If the presetnation is not delay during playback this delay is used to calculate the presentation timestamps.

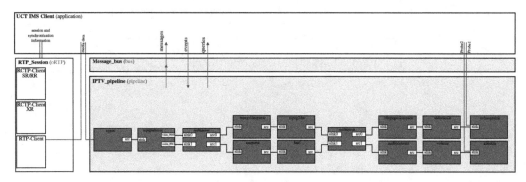

Figure 5.39: GStreamer pipeline for decoding and presentation

## 5.4.2 Callback function for starting IPTV-Session (sc_start_watching)

The function shown in figure 5.40 is used to start the threads for the IPTV session. This function is called from the GUI implemented into the UCT IMS Client. The function **playIPTV()** is used to call **initialiseIptvVideoPipeline()** with the pointer to the running SIP session.

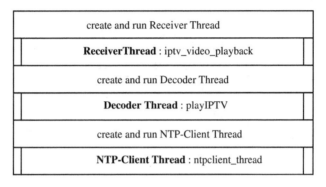

Figure 5.40: Callback function for starting the IPTV functions

## 5.4.3 Callback function for terminating IPTV-Session (sc_stop_watching)

The function shown in figure 5.41 is used to stop the threads for the IPTV session. This function is called from the GUI implemented into the UCT IMS Client.

| stop Receiver Thread |
| --- |
| free memory used by Receiver Thread |
| stop Decoder Thread |
| free memory used by Decoder Thread |
| stop NTP-Client Thread |
| free memory used by NTP-Client Thread |

Figure 5.41: Callback function for stoping the IPTV functions

### 5.4.4 Function for starting the Decoder (initialiseIptvVideoPipeline)

The GStreamer pipeline, described in section 5.4.1 for watching synchronized IPTV is created by using the function shown in figure 5.42. The infinite loop is stop by termianting the thread using the callback function from the GUI (section 5.4.3).

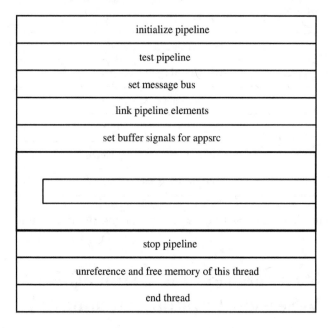

Figure 5.42: Function for starting the Decoder

## 5.4.5 RTP receiver (iptv_video_playback)

The RTP session for viewing IMS based IPTV is created with **iptv_video_playback()**
(figure 5.43). During the runtime the received XR IDMS messages are used to calculate
the presentation delay. If the delay less then 15 s the GStreamer pipeline is paused and
the calculated delay is added to the presentation delay.

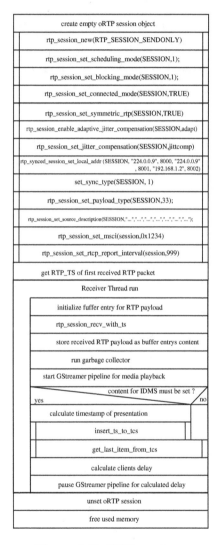

Figure 5.43: RTP receiver

## 5.5 SC application on provider side

If the UE does not support the synchronization flow of IMS based IPTV, the SC has to be implemented on provider side to support synchronization. This section will give an overview of such an application.

### 5.5.1 RTP receiver (rtp_receiver_run)

Figure 5.44 shows the creation of the RTP receiver for the provider side SC. During runtime IDMS reports are created and the received RTP stream is passed to a buffer for sending.

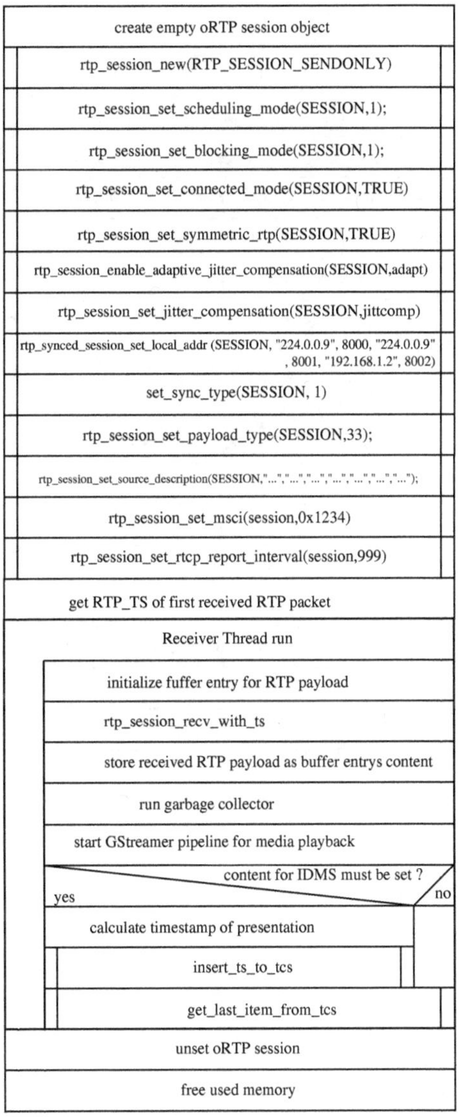

| |
|---|
| create empty oRTP session object |
| rtp_session_new(RTP_SESSION_SENDONLY) |
| rtp_session_set_scheduling_mode(SESSION,1); |
| rtp_session_set_blocking_mode(SESSION,1); |
| rtp_session_set_connected_mode(SESSION,TRUE) |
| rtp_session_set_symmetric_rtp(SESSION,TRUE) |
| rtp_session_enable_adaptive_jitter_compensation(SESSION,adapt) |
| rtp_session_set_jitter_compensation(SESSION,jittcomp) |
| rtp_synced_session_set_local_addr (SESSION, "224.0.0.9", 8000, "224.0.0.9", 8001, "192.168.1.2", 8002) |
| set_sync_type(SESSION, 1) |
| rtp_session_set_payload_type(SESSION,33); |
| rtp_session_set_source_description(SESSION,"...","...","...","...","...","...","...","..."); |
| rtp_session_set_msci(session,0x1234) |
| rtp_session_set_rtcp_report_interval(session,999) |
| get RTP_TS of first received RTP packet |
| Receiver Thread run |
| initialize fuffer entry for RTP payload |
| rtp_session_recv_with_ts |
| store received RTP payload as buffer entrys content |
| run garbage collector |
| start GStreamer pipeline for media playback |
| content for IDMS must be set ? yes / no |
| calculate timestamp of presentation |
| insert_ts_to_tcs |
| get_last_item_from_tcs |
| unset oRTP session |
| free used memory |

Figure 5.44: RTP receiver of provider side SC

## 5.5.2 RTP sender (rtp_sender_run)

The RTP sender of the provider side SC is shown in figure 5.45. The buffer from the RTP receiver is sent to the UE using a unicast RTP session.

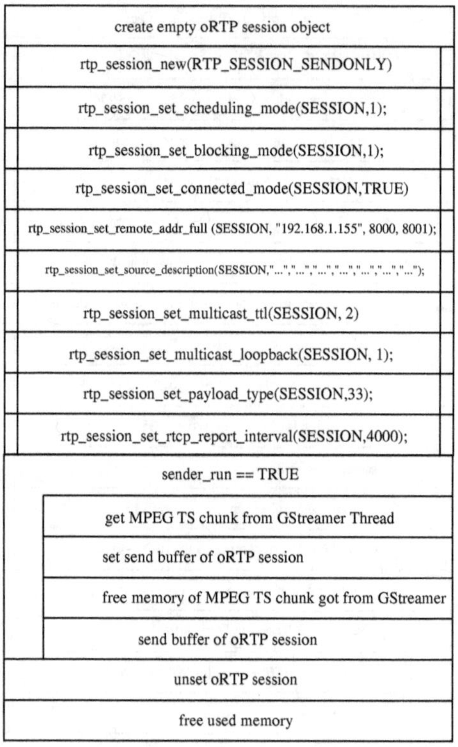

Figure 5.45: RTP sender of provider side SC

### 5.5.3 Graphical user interface (gui_run)

The MDF's usage is mainly in a non graphical environment, which makes it necessary to use a console based output for status information. Figure 5.46 shows the thread function for the graphical console output. The output is formatted by using ncurses[17]. By pressing F9 the user is able to exit the application.

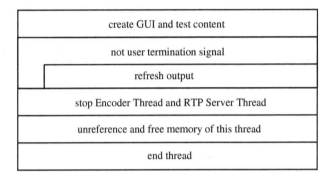

Figure 5.46: Provider side SC - GUI thread

---

[17]The usage of this library is explained by [50].

## 5.5.4 Primary function of the application (main)

Figure 5.47 shows the main function of the application. This function is started as default by running the application and initializes and runs the threads for the sender, the receiver and the GUI.

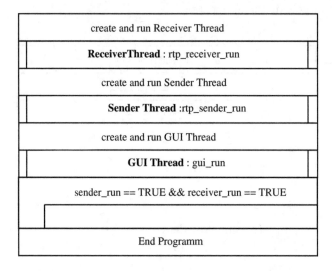

Figure 5.47: Main fucntion of provider side SC

## 5.6 Transcoder application

In a flexible IPTV environment it is necessary to transcode the media stream to a compatible format for the UE. This is mostly used for mobile device, which are mostly not able to receive media streams in the same quality as Set Top Boxes.

### 5.6.1 GStreamer Pipeline

The GStreamer pipeline shown in figure 5.48 is a simple pipeline for converting the video stream to a lower resolution. The received stream has the same stream format as the sending stream.

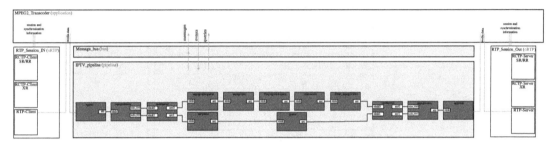

Figure 5.48: Gstreamer pipeline of the transcoder

## 5.6.2 RTP receiver (rtp_receiver_run)

Figure 5.49 shows the creation of the RTP receiver for the transcoder. During run-time IDMS reports are created and the received RTP stream is passed to a buffer for sending.

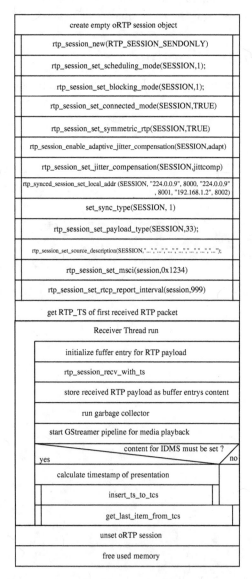

Figure 5.49: RTP receiver of the transcoder

## 5.6.3 RTP sender (rtp_sender_run)

The RTP sender of the transcoder is shown in figure 5.50. The buffer from the RTP receiver is sent to the UE using a unicast RTP session.

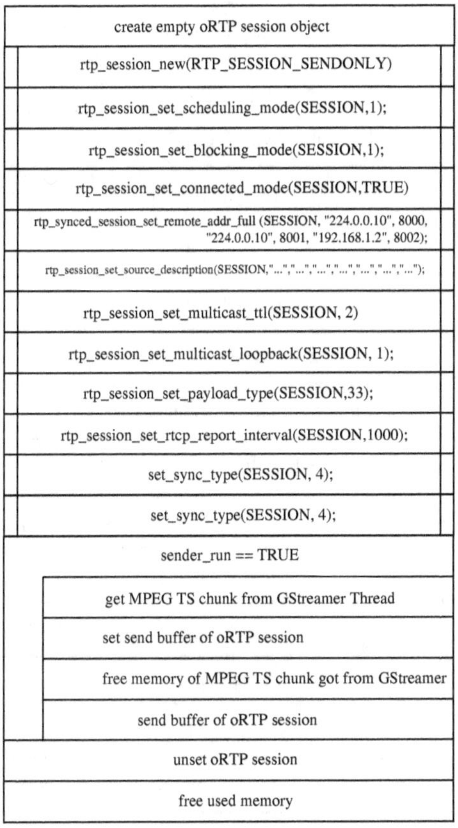

| create empty oRTP session object |
|---|
| rtp_session_new(RTP_SESSION_SENDONLY) |
| rtp_session_set_scheduling_mode(SESSION,1); |
| rtp_session_set_blocking_mode(SESSION,1); |
| rtp_session_set_connected_mode(SESSION,TRUE) |
| rtp_synced_session_set_remote_addr_full (SESSION, "224.0.0.10", 8000, "224.0.0.10", 8001, "192.168.1.2", 8002); |
| rtp_session_set_source_description(SESSION,"...","...","...","...","...","...","..."); |
| rtp_session_set_multicast_ttl(SESSION, 2) |
| rtp_session_set_multicast_loopback(SESSION, 1); |
| rtp_session_set_payload_type(SESSION,33); |
| rtp_session_set_rtcp_report_interval(SESSION,1000); |
| set_sync_type(SESSION, 4); |
| set_sync_type(SESSION, 4); |
| sender_run == TRUE |
| get MPEG TS chunk from GStreamer Thread |
| set send buffer of oRTP session |
| free memory of MPEG TS chunk got from GStreamer |
| send buffer of oRTP session |
| unset oRTP session |
| free used memory |

Figure 5.50: RTP sender of the transcoder

## 5.6.4 Media transcoder (transcoder_thread_run)

The GStreamer pipeline for the Transcoder is created with the function shown in figure 5.51.

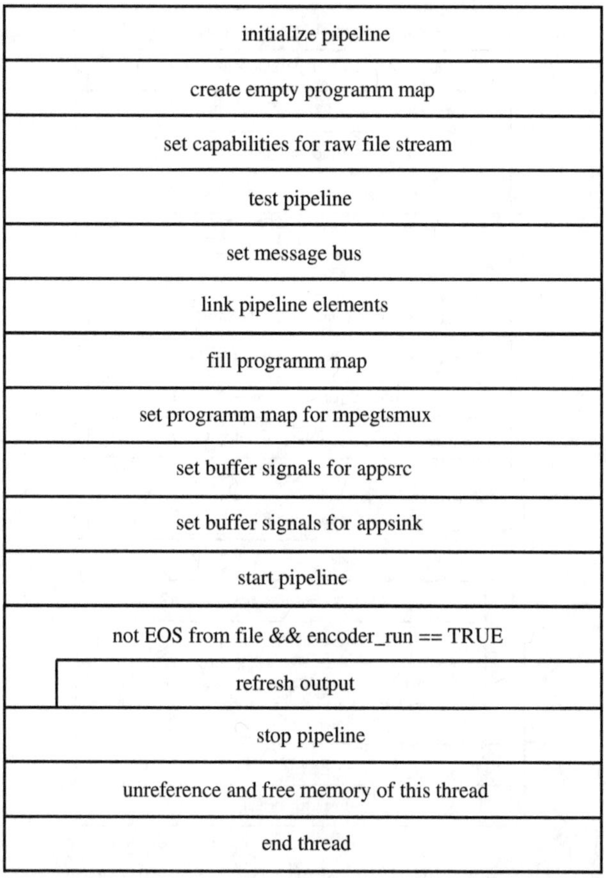

| |
|---|
| initialize pipeline |
| create empty programm map |
| set capabilities for raw file stream |
| test pipeline |
| set message bus |
| link pipeline elements |
| fill programm map |
| set programm map for mpegtsmux |
| set buffer signals for appsrc |
| set buffer signals for appsink |
| start pipeline |
| not EOS from file && encoder_run == TRUE |
| refresh output |
| stop pipeline |
| unreference and free memory of this thread |
| end thread |

Figure 5.51: Transcoder function of the transcoder application

## 5.6.5 Graphical user interface (gui_run)

The MDF's usage is mainly in a non graphical environment, which makes it necessary to use a console based output for status information. Figure 5.52 shows the thread function for the graphical console output. The output is formatted by using ncurses[18]. By pressing F9 the user is able to exit the application.

---

[18]The usage of this library is explained by [50].

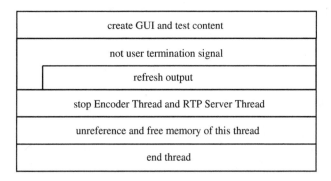

Figure 5.52: Transcoder - GUI thread

## 5.6.6 Primary function of the application (main)

Figure 5.53 shows the main function of the application. This function is started as default by running the application and initializes and runs the threads for the sender, the receiver, the media Transcoder and the GUI.

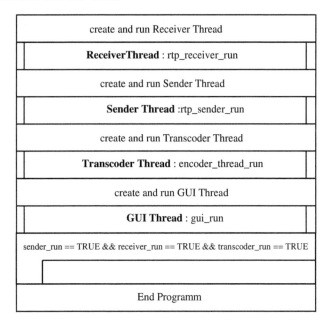

Figure 5.53: Main function of the transcoder

## 5.7 Multistream source transcoder

The Transcoder described in section 5.6 could be modified to serve the needs of a synchronized Transcoder, which mixes multiple input streams to one output stream. Figure 5.54 shows the GStreamer pipeline fo such a Transcoder. This pipeline is a modification of the pipeline shown in figure 5.48 by using the GStreamer example found at [51].

For a application the functions described in section 5.6 could be used by instantiating the receiver thread, if all senders use the same RTP_TS for the RTP packets, which belong to each other. Each instance of the receiver should have its own buffer to the GStreamer thread, which sorts the content during passing it to the pipeline using the *appsrc* modules shown in figure 5.54.

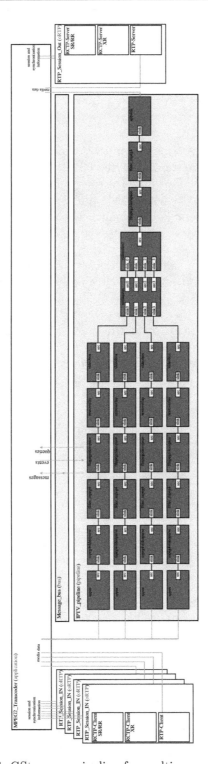

Figure 5.54: GStreamer pipeline for multisource transcoder

# 6 Evaluation

The IDMS implementation is evaluated in the following described steps. First of all the library has to be evaluated to show, that the RTCP report blocks and messages are created correctly and the timestamps have the correct format. If this examination is passed the applications are tested to show that they run in several instances on the same system (UCT IMS Client), they are able to communicate the way they should and finally the desired synchronization could be done using the applications.

For the protocol analysis a modified Wireshark V1.4.4 (see [52]), which interprets the values of the XR IDMS report is used. Source code, description and sample capture file are available at [53].

The playout difference between two video outputs is measured using a measurement system introduced by *Automatic Measurement of Play-out Differences for Social TV, Interactive TV, Gaming and Inter-destination Synchronization* [54]. For the measurement a video with both receivers is analyzed by using the cross-correlation of the scene changes of both video outputs.

The figures in that section are created using the following scripts, stored on the additional DVD

- *measurement_results/plots/plot_res.m* - for plotting the results of the visual measurement,

- *measurement_results/plots/plots/plot_it.gnuplot* - for plotting the output of the SC and MSAS and

- *measurement_results/plots/plots/convert_ps_plots.sh* - for conversion of gnuplots output.

The plots of the SC and MSAS output is done with the following command set.

```
gnuplot -e plot_it.gnuplot && sh convert_ps_plots.sh
```

# 6.1 Evaluation of the protocol implementation

For the protocol evaluation a dumb RTP sender was created, which sends content of a file to a IP multicast group. This is a modification[19] of the *rtpsend* example from the linphone project [45], which is included in the sources of oRTP. The RTP receiver is a modified[20] version of the *rtptcv* example from the linphone project [45]. A combination of both makes the Transcoder which also send only dumb RTP packets. The MSAS was represented by an early release, which only reflects received packets to the sender, without modification. In figure 6.1 the outputs of Wireshark are shown. Each report is interpreted

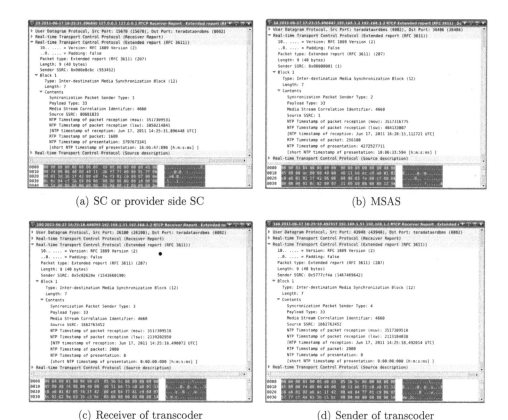

(a) SC or provider side SC

(b) MSAS

(c) Receiver of transcoder

(d) Sender of transcoder

Figure 6.1: XR IDMS reports measured with Wireshark

the right way, which shows, that the library is able to send correct IDMS XR reports using

---

[19]The example is a unicast RTP sender.

[20]The function **rtp_synced_session_set_local_addr()** instead of **rtp_session_set_local_addr()** is used, a multicast receiver is created and the MSAS address is set.

RTCP on applications request.For the analysis of synchronization relevant RTCP flows the filters shown in table 6.1 are useful.

| Filter name | Description |
| --- | --- |
| rtcp.xr.idms.msci == NNN | filters all packages by MSCI value equals NNN |
| rtcp.xr.idms.source_ssrc == NNN | filters all packages by source SSRC value equals NNN |
| rtcp.xr.idms.spst == NNN | filters all packages by SPST value equals NNN |

Table 6.1: Useful Wireshark filters for IDMS XR reports

## 6.2 Evaluation of the applications

The first step of the applications evaluation is testing the calculation used by the MSAS. Followed by testing the instantiation of library, by using two instances of the client on one *Personal Computer* (PC) without virtualization. This will show that the library could be used for a Trancoder like the one decribed in section 5.7 on page 98 and applications using this library are able to interact on one PC without influencing each other. For interpretation of the performance of the hole system two reference measurements (one with another application and the other with the designed SC without enabled synchronization) are compared to measurements with clients trying to synchronize by communicating to the MSAS.

## 6.2.1 Timestamp estimation

The MSAS application stores the calculation results and received timestamps in a *Comma-Separated Values* (CSV) file. With gnuplot (see [55] for detailed description) these files could be shown as figures like figure 6.2 or figure 6.3. For this test two SC, on two PC are communicating to the MSAS, which generate the values for the plot.

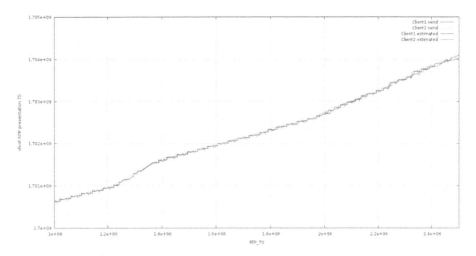

Figure 6.2: Estimation of the received presentation timestamps

Figure 6.2 shows the results of the test over the time of testing. These results are zoomed in figure 6.3, to show the estimation error between the received and the estimated timestamps. Goal of the synchronization system is to keep the delay of play-out lower then one second, which leeds to a maximal estimation error of less then 0.5 seconds if the clocks are perfekt synchronized. In a practical environment the error schould be less then 0.1 second.

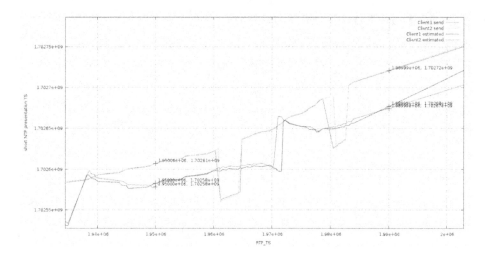

Figure 6.3: Estimation of the received presentation timestamps (zoomed)

The values used for plotting the figures are used to obtain the values of table 6.2. These values goat on the one side, that the error is too high for the synchronization, but on the other side the difference between both clients presentation timestamp is approximately the same as the difference between both estimated functions. This only shows, that using a linear estimation can solve the problem of calculating the presentation delay at the MSAS.

| Value | Client 1 [s] | Client 2 [s] |
|-------|--------------|--------------|
| min   | 0,0001068132 | 0,0011139086 |
| max   | 2,2820935378 | 1,9638361181 |
| avg   | 0,5682614394 | 0,4231202605 |

Table 6.2: Absolute error between received timestamps and estimated timestamps

## 6.2.2 Measuring using one PC

For a stable software it is necessary to show, that library and application are able to run in multiple instances on one PC. Figure 6.4 shows the test set with running instances of the SC application. This test set is used to determine the play-out delay between this clients

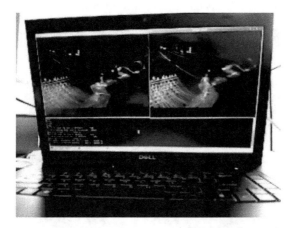

Figure 6.4: Test set for proving the instantiation of library and client

using the measurement system described by [54]. The results are shown in figure 6.5, which shows a delay of -32 frames (shown in 6.5(b)). Because of the frame rate of 30 frames per second the delay is -1.066 s, which means the left instance is 1.066 seconds ahead of the right one.

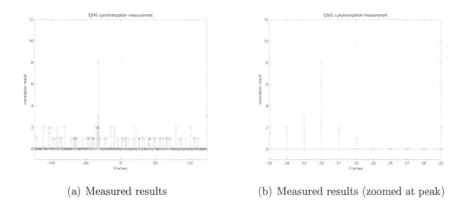

(a) Measured results       (b) Measured results (zoomed at peak)

Figure 6.5: Measurement results for two instances of the SC

### 6.2.3 Measuring using two PC

After proof of the instantiation, reference values of unsynchronized clients have to be known to compare these to measurements in a synchronized environment. In figure 6.6 the test set for all measurements using two PC are shown. The video output scaled approximately to same sizes for each test, to lower error.

Figure 6.6: Test set using two PC for measuring

The result of the measurement is chown in figure 6.7, the delay is -1.2 s. This result is one of the reference values for the comparison to the synchronized environment.

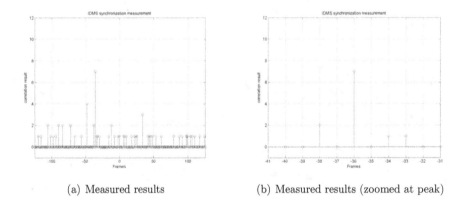

(a) Measured results      (b) Measured results (zoomed at peak)

Figure 6.7: Measurement without MSAS on two PC

## 6.2.4 Measuring between two clients using VLC

The measurement set described in the last section by figure 6.6 is used to get the last two reference values for a comparison. First VLC is started at both PC. In the second test on the left PC UCT IMS client is started and on the right PC VLC.

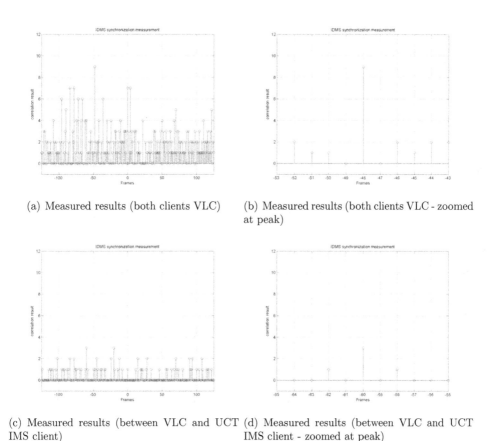

(a) Measured results (both clients VLC)

(b) Measured results (both clients VLC - zoomed at peak)

(c) Measured results (between VLC and UCT IMS client)

(d) Measured results (between VLC and UCT IMS client - zoomed at peak)

Figure 6.8: Reference measurement with VLC on two PC

The results of the measurement are shown in figure 6.8. Figures 6.8(a) and 6.8(b) show the results for the measurement using VLC on both PC and figures 6.8(c) and 6.8(d) show the measurement using UCT IMS client and VLC. Peaks in measurement results are at -1.6 s, for both clients are VLC and -2.0 for the other measurement. In 6.8(c) the disturbance is very high. The reason for that is, that the delay is more then 4 seconds and out of range of the measurement system in the normal configuration. The video is stored on the additional DVD in *measurement_results/uct_vlc.wmv* and shows that the delay between the play-out of both application is more then 2 s. For the next steps only the delay of the first measurement between twice VLC is used.

## 6.2.5 Measuring the IDMS implementation

### 6.2.5.1 Measurment with enabled pausing

The first measurement with two SC connected to the MSAS is done with enabled pausing. Clients tried to synchronized using the advices given by the MSAS. Results of the visual measurement are shown in 6.9(a) with the peak shown in 6.9(b).

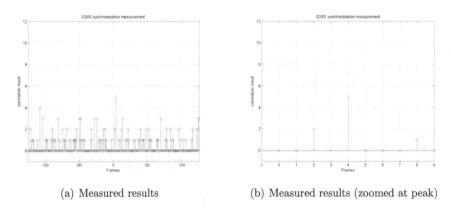

(a) Measured results  (b) Measured results (zoomed at peak)

Figure 6.9: Measurement with MSAS on two PC - pausing enabled

The result of 133 ms delay between the clients would be a very good results for the synchronization algorithm, but this value has a high error depending on the measurement algorithm. For calculation of this result a cross-correlation between the scene changes of both video outputs is done. This cross-correlation needs a nearly fixed distance of scene changes in both videos to get sufficient results. The pausing, which is done by both clients lead to an inaccuracy in this distance, which is also represented by the large amount of disturbance, shown in 6.9(a).

| Value | Left client[21] | Right client[22] |
|---|---|---|
| MSAS error min | 6.1036E-005 | 6.103608E-005 |
| MSAS error max | 3.9989318 | 4.1412069886 |
| MSAS error avg | 0.507631836 | 0.9248710249 |
| Client delay min | 0 | 0 |
| Client delay max | 4.50779 | 3.979217 |
| Client delay avg | 0.17249 | 0.167108 |
| visual measure | 0.133 | |

Table 6.3: Measurement results using two clients with enabled pausing

The absolute error and the measured delays at both clients are shown in table 6.3 in comparison to the measurement with the visual measurement using the system decribed by [54]. The high estimation error, shown in 6.10(f), leads to a high error in delay calculation. The results of the clients calculation is useless, because of this high error. At this point the system should only be used to show, that it is possible to synchronize and to give advices on how to implement the communication using IDMS reports in RTCP XR messages.

The video containing this measurement is located on the additional DVD to this book (location: *measurement_results/ref1.wmv*). This video shows, that even if the results will be good the pausing has to be improved, because the output pauses that often that normal viewing is not possible. The video also shows, that the resulting delay is more then 20 s, because the pausign is done that often, the the estiamtion at the MSAS got bad. Figure 6.10 shows all results of that measurement. In the zoomed views the steps in the presentation timestamps are visible, which causes the high error in the estimation.

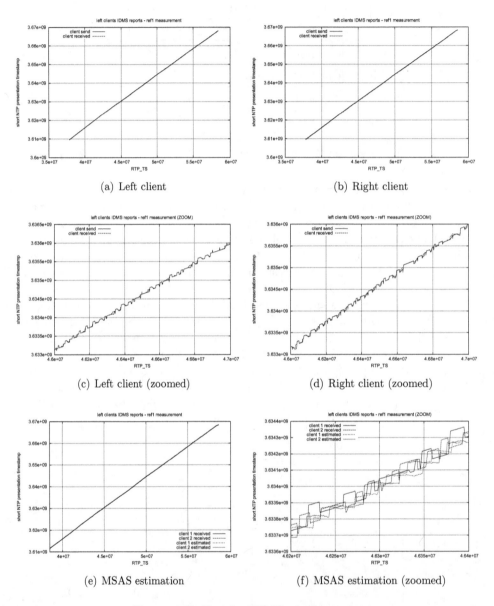

(a) Left client

(b) Right client

(c) Left client (zoomed)

(d) Right client (zoomed)

(e) MSAS estimation

(f) MSAS estimation (zoomed)

Figure 6.10: Results IDMS measurement 1

### 6.2.5.2 Measurment with disabled pausing

The previous measurements results have to be proved, by an other measurement. For that purpose, the pausing at the clients was disabled. If the synchronization algorithm is working the calculated delays should have one peak at the maximum delay (a good algorithm will have this point at the beginning) and after pausing the calculated delays should be less then that. During that measurement a video is made, because the visual measurement will deliver precise results if the clients are not pausing.

The results of this measurement is shown in table 6.4. The calculated delays at the clients are not useful for any synchronization purpose.

| Value | Left client[23] | Right client[24] |
|---|---|---|
| MSAS min | 0.000030518 | 9.155E-005 |
| MSAS max | 2.9881132 | 4.1986266 |
| MSAS avg | 0.537151 | 0.5707611 |
| client min | 0 | 0 |
| client max | 3.56198 | 2.62117004 |
| client avg | 0.26714 | 0.2330245145 |
| optical measure | -2.467 | |

Table 6.4: Measurement results using two clients with dissabled pausing

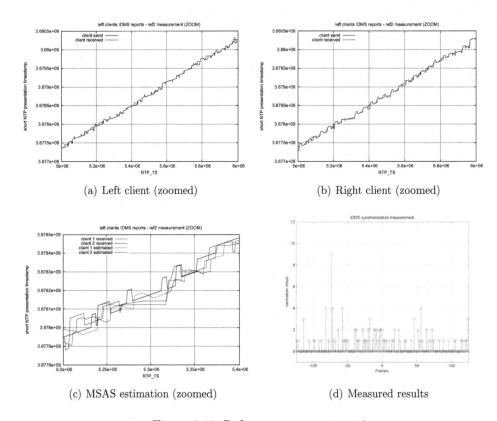

(a) Left client (zoomed)  (b) Right client (zoomed)

(c) MSAS estimation (zoomed)  (d) Measured results

Figure 6.11: Reference measurement 2

Figure 6.11 shows the plotted measurement results. The steps in presentation timestamp incrementation, which causes the high error are shown in 6.11(a), 6.11(b) and 6.11(c). Figure 6.11(d) shows the graphical output of the play-out delay measurement, which is even worse then the results using unsynchronized VLC clients (see section 6.2.4 on page 107). The other plots are stored on the additional DVD in the folder *measurement_results/plots*.

### 6.2.5.3 Synchronization of two SC on one PC

The last two evaluation tests showed, that steps in the time incrementation causes high error during estimation of the linear function in the MSAS. This has to be proofed by disabling the incrementation in decoding delay in the client and using the same PC for both clients. In the last test the client only stopped pausing, but sends corrected (by incrementing the decoding delay) presentation times back to the MSAS. Each time such a correction is made it causes a step in the timestamp increment over time. The second influence in the time stamping is the NTP client of the SC, which sets an offset to the local clock after synchronizing to a NTP server. The following two test sets use two SC

| Value | Measurement 3 | | Measurement 4 | |
|---|---|---|---|---|
| | Left client | Right client | Left client | Right client |
| MSAS min | 0.0004730297 | 0.0001678492 | 0 | 7.62951094834821E-005 |
| MSAS max | 0.9730830854 | 0.8758983749 | 0.8693064775 | 1.5058365759 |
| MSAS avg | 0.4343028801 | 0.4061786289 | 0.3995372918 | 0.4392763658 |
| client min | 0 | 0 | 1.10484314 | 0 |
| client max | 1.57746887 | 0 | 2.4077301 | 0 |
| client avg | 0.7422347549 | 0 | 1.7712924555 | 0 |
| optical measure | -0.273 | | -3.3 | |

Table 6.5: Results of measurement 3 and 4

on one PC synchronizing by communication to an MSAS. At this point only the interesting figures are shown, a complete set is stored on the additional DVD in the folder *measurement_results/plots*.

The results of both measurements are shown in table 6.5. With the modification in the test set, the performance of the synchronization algorithm has improved. That fact leads to the summary, that by improving the estimation of the MSAS or by modifying the hole synchronization algorithm this system is able to synchronize the play-out of several SC. The modification of the algorithm could be, that all clients have to send XR IDMS reports on received SR, which will make the estimation unnecessarily.

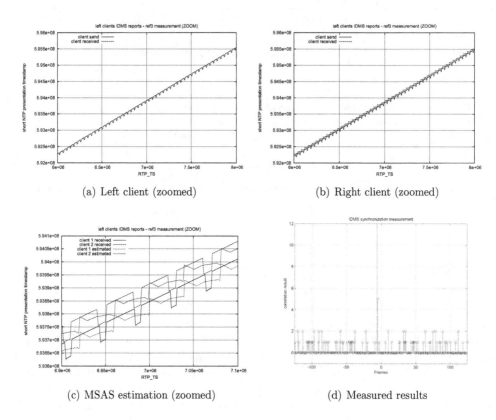

(a) Left client (zoomed)

(b) Right client (zoomed)

(c) MSAS estimation (zoomed)

(d) Measured results

Figure 6.12: Reference measurement 3

Figures 6.12 and 6.13 show the specific difference to the results of the last two measurement. The most interesting results are shown in 6.12(c) and 6.13(c). The difference between received timestamps is approximately the same as between the estimated timestamps of both clients. The resulting error in the delay calculation of the clients is caused in the error between received timestamp and estimated timestamp of the client with the latest presentation. In case of 6.12(c) it is client $2$[25] and in case of 6.13(c) it is client $1$[26].

---

[25]In the test set this is the right client.
[26]In the test set this is the right client.

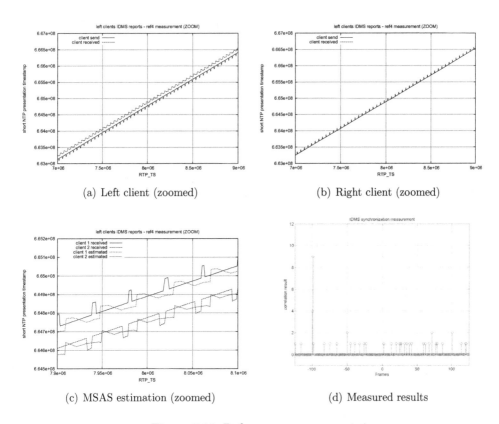

(a) Left client (zoomed)

(b) Right client (zoomed)

(c) MSAS estimation (zoomed)

(d) Measured results

Figure 6.13: Reference measurement 4

Both measurement results show, that there is a problem in the timestamp generation, which is depending either on the NTP clients offset, the calculation of the presentation timestamp in the SC application using the measurement decoding delay or both. For a working environment, this has to be fixed.

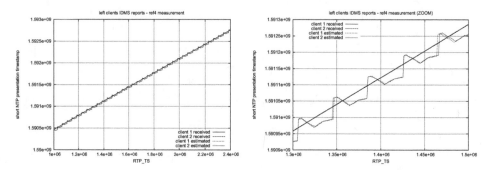

(a) Error in estimation with disabled NTP client (b) Error in estimation with disabled NTP client (zoomed)

Figure 6.14: Estimation Error - disabled NTP client

The NTP client of both clients was disabled before the last measurement was done. The resulting output of the MSAS is shown in figure 6.14. The estimation error is still too high for good results, but now it is only depended on the averaging done for the fractional part of the timestamp.

# 7 Conclusion and Future Work

## 7.1 The IDMS implementation

In the last chapters a prototyped implementation for IDMS in IMS based IPTV using ETSI TS 183 063 Annex W is described. Results shown in chapter 6 show, that it is possible to synchronize RTP recievfers using this system, but also show that there has to be done more on the applications used for prototyping. The main goal of this research, implementing the protocol extension for IDMS in IMS based IPTV are done by providing a modified version of oRTP, which enables applications to communicate using RTCP XR blocks for IDMS. The results also show that it is possible to use a MSAS for sending synchronization advices and that SC are able to recieve them and to calculate their delay. In the theoretical research on that topic some advices on how to improve the results, by using a NTP client inside of the SC or by measuring the delay between SC and MSAS are given. The next two sections will give some additional recomandations, which will improve interoperability between several implementations of the stadard, followed by some extension for the implementation based on this thesis.

## 7.2 Recomandations for the protocol description

In ETSI TS 182 027 [2] and ETSI TS 183 063 [2] are some non specific advices on the implementation given which could lead to problems in the interoperability between different implementations of these standards. The following two recommendations should be made by either ETSI TS 182 027 [2] or ETSI TS 183 063 [2] to avoid bad results.

1. MSAS shall send IDMS instructions depending on the same RTP_TS as the SC has send in the last XR IDMS report.

2. To avoid high clock drifts, which will decrease the result of the synchronization minimal interval of clock synchronization should be given.

3. For a good result it should be recommended, that clients send XR IDMS reports on received SR (report on the same RTP packet) or that the MSAS has to do a sufficient[27] estimation.

---

[27] At this point a maximum error could not be given. This has to be topic of another research.

4. Clients should support delay measurement between non-senders as described by RFC3611[20]. This will support indication of unsynchronized clocks.

# 7.3 Possible Extensions

## 7.3.1 Library

1. Modification of the XR report parser to use the signaling system of oRTP.

2. Implementation of the RTT and DLRR RTCP XR report blocks for measuring the delay between SC and MSAS on applications request and passing results to the application.

3. Creation of all IMS based IPTV relevant Payload Types and their properties.

4. Modification of the RTCP parser to support reports send by Transcoders to fill a list of RTP_TS references for received in comparison to sent RTP packets.

## 7.3.2 Client

1. The calculation of the presentation timestamp has to be improved.

2. For a better QoE video should be played faster or slower to synchronize then pausing the play-out.

## 7.3.3 RTP-Sender

1. DVB stream as a source of the RTP sender, which will improve interoperability with existing systems.

2. A MCF could be implemented by using GStreamers *Real Time Streaming Protocol* (RTSP) plugin.

## 7.3.4 Transcoder

1. XML import for dynamical configuration of the output stream of the Transcoder by using a modified GStremer XML import.

2. Configuration interface for setting parameters of the RTP sender and the RTP receiver.

### 7.3.5 MSAS

1. Improvement of the estimation of the linear presentation function.

2. step detection to improve the linearization.

3. communication to the session based MSAS for creating multiple calculation instances (one for each MSCI) for a stream.

## 7.4 Research questions for future work

The implementation described in this thesis supports only the basic functionality needed. This leads to the following possible research questions for future work.

1. Impacts on the synchronization depending on the used algorithm using different implementations on the client or MSAS side. For example clients which are only pausing and clients which only modify the speed of playing the content in a mixed environment.

2. How could IMS based IPTV synchronized to a local DVB stream? TV streams, which have a high potential on togetherness by the audience, are mostly used by several TV systems like DVB-T, DVB-C or DVB-S. In future deployments of TV infrastructure synchronization of such systems with an IPTV system could lead to better quality, the users will recognize.

# Bibliography

[1] TISPAN, "ETSI TS 182 063 V3.5.2: IPTV IMS scope stage 3," Mar 2011.

[2] TISPAN, "ETSI TS 182 027 V3.5.1: IPTV Architecture; IPTV functions supported by the IMS subsystem," Mar. 2011.

[3] "IEEE 802.3-2008 ieee standard for information technology–telecommunications and information exchange between systems–local and metropolitan area networks–specific requirements part 3: Carrier sense multiple access with collision detection (csma/cd) access method and physical layer specifications - section one," 12 2008.

[4] D. Geerts, I. Vaishnavi, R. Mekuria, O. van Deventer, and P. César, "Are we in sync?: synchronization requirements for watching online video together." in CHI, D. S. Tan, S. Amershi, B. Begole, W. A. Kellogg, and M. Tungare, Eds. ACM, 2011, pp. 311–314. [Online]. Available: http://dblp.uni-trier.de/db/conf/chi/chi2011.html#GeertsVMDC11

[5] "IEEE 1588-2008 ieee standard for a precision clock synchronization protocol for networked measurement and control systems," 7 2008.

[6] D. Mills, "Network Time Protocol (Version 3) Specification, Implementation and Analysis," RFC 1305 (Draft Standard), Mar. 1992. [Online]. Available: http://www.ietf.org/rfc/rfc1305.txt

[7] J. Rosenberg, H. Schulzrinne, G. Camarillo, A. Johnston, J. Peterson, R. Sparks, M. Handley, and E. Schooler, "SIP: Session Initiation Protocol," RFC 3261 (Proposed Standard), Jun. 2002, updated by RFCs 3265, 3853, 4320, 4916. [Online]. Available: http://www.ietf.org/rfc/rfc3261.txt

[8] J. Rosenberg and H. Schulzrinne, "Reliability of Provisional Responses in Session Initiation Protocol (SIP)," RFC 3262 (Proposed Standard), Jun. 2002. [Online]. Available: http://www.ietf.org/rfc/rfc3262.txt

[9] J. Rosenberg and H. Schulzrinne, "Session Initiation Protocol (SIP): Locating SIP Servers," RFC 3263 (Proposed Standard), Jun. 2002. [Online]. Available: http://www.ietf.org/rfc/rfc3263.txt

[10] J. Rosenberg and H. Schulzrinne, "An Offer/Answer Model with Session Description Protocol (SDP)," RFC 3264 (Proposed Standard), 2002. [Online]. Available: http://www.ietf.org/rfc/rfc3264.txt

# Bibliography

[11] A. B. Roach, "Session Initiation Protocol (SIP)-Specific Event Notification," RFC 3265 (Proposed Standard), Jun. 2002. [Online]. Available: http://www.ietf.org/rfc/rfc3265.txt

[12] M. Handley, V. Jacobson, and C. Perkins, "SDP: Session Description Protocol," RFC 4566 (Proposed Standard), Jul. 2006. [Online]. Available: http://www.ietf.org/rfc/rfc4566.txt

[13] F. Andreasen, "Session Description Protocol (SDP) Simple Capability Declaration," RFC 3407 (Proposed Standard), Oct. 2002. [Online]. Available: http://www.ietf.org/rfc/rfc3407.txt

[14] H. Schulzrinne, S. Casner, R. Frederick, and V. Jacobson, "RTP: A Transport Protocol for Real-Time Applications," RFC 3550 (Standard), Jul. 2003. [Online]. Available: http://www.ietf.org/rfc/rfc3550.txt

[15] G. Siegmund, Technik der Netze: Neue Ansätze: SIP in IMS und NGN, 6th ed. Hüthig, 10 2009. [Online]. Available: http://amazon.de/o/ASIN/3778540637/

[16] JTC-DVB, "ETSI EN 300 429 V1.2.1: Framing structure, channel coding and modulation for cable systems DVB-C," Apr 1998.

[17] J. Postel, "User Datagram Protocol," RFC 768 (Standard), Aug. 1980. [Online]. Available: http://www.ietf.org/rfc/rfc768.txt

[18] L. Boudec, J.Y., and P. Thiran, Network calculus: a theory of deterministic queuing systems for the internet. springer-Verlag, 2001.

[19] Z. Albanna, K. Almeroth, D. Meyer, and M. Schipper, "IANA Guidelines for IPv4 Multicast Address Assignments," RFC 3171 (Best Current Practice), Aug. 2001. [Online]. Available: http://www.ietf.org/rfc/rfc3171.txt

[20] T. Friedman, R. Caceres, and A. Clark, "RTP Control Protocol Extended Reports (RTCP XR)," RFC 3611 (Proposed Standard), Nov. 2003. [Online]. Available: http://www.ietf.org/rfc/rfc3611.txt

[21] A. Pathak, H. Pucha, Y. Zhang, Y. Hu, and Z. Mao, "A measurement study of internet delay asymmetry," in Proceedings of the 9th international conference on Passive and active network measurement. Springer-Verlag, 2008, pp. 182–191.

[22] J. Ott, J. Chesterfield, and E. Schooler, "RTP Control Protocol (RTCP) Extensions for Single-Source Multicast Sessions with Unicast Feedback," RFC 5760 (Proposed Standard), Feb. 2010, updated by RFC 6128. [Online]. Available: http://www.ietf.org/rfc/rfc5760.txt

[23] B. Huntgeburth, M. Maruschke, and S. Schuhmann, "Open-Source Based Prototype of Quality of Service (QoS) Monitoring and Quality of Experience (QoE) Estimation in Telecommunication Environments," 2011.

[24] 3GPP, "3GPP TS 23.228 V10.5.0 : IP Multimedia Subsystem (IMS); Stage 2," Jun 2011.

[25] TISPAN, "ETSI ES 282 001 V3.4.1: TISPAN NGN Release 3 architecture," Sep 2009.

[26] TISPAN, "ETSI TS 123 228 V10.5.0 : IP Multimedia Subsystem (IMS); Stage 2," Jun 2011.

[27] Content: HfT Leipzig; Plattform: Dominik Richter, "NGN Lernmodul HfT Leipzig," date last viewed: 15-07-2011. [Online]. Available: http: //www.hft-leipzig.de/maruschke/lernmodul2010/start.html

[28] TISPAN, "ETSI TS 123 002 V10.2.0: Digital cellular telecommunications system (Phase 2+); Universal Mobile Telecommunications System (UMTS); LTE; Network architecture (3GPP TS 23.002 version 10.2.0 Release 10) ," Mar 2011.

[29] J. Rosenberg, "The Session Initiation Protocol (SIP) UPDATE Method," RFC 3311 (Proposed Standard), Oct. 2002.

[30] 3GPP, "3GPP2 X.S0013-009-0 V1.0 : IMS/MMD Call Flow Examples," Jun 2011.

[31] F. Boronat, J. Lloret, and M. García, "Multimedia group and inter-stream synchronization techniques: A comparative study," Information Systems, vol. 34, no. 1, pp. 108–131, 2009.

[32] I. Openmoko, "Openmoko developer guide - Openmoko," date last viewed: 15-07-2011. [Online]. Available: http://wiki.openmoko.org/wiki/Openmoko_developer_guide

[33] "BeagleBoard.org," date last viewed: 15-07-2011. [Online]. Available: http: //pandaboard.org/

[34] "Pandaboard," date last viewed: 15-07-2011. [Online]. Available: http: //beagleboard.org/

[35] Fraunhofer FOKUS, "OpenIMS," date last viewed: 15-07-2011. [Online]. Available: http://www.openimscore.org/

[36] The Apache Software Foundation, "Tomcat," date last viewed: 15-07-2011. [Online]. Available: http://tomcat.apache.org/index.html

[37] Oracle, "Glassfish," date last viewed: 15-07-2011. [Online]. Available: http: //glassfish.java.net/

[38] Oracle Corporation and/or its affiliates, "JSR-000289 SIP Servlet 1.1 (Final Release)," date last viewed: 03-04-2011. [Online]. Available: http://jcp.org/aboutJava/communityprocess/final/jsr289/index.html

# Bibliography

[39] Oracle Corporation and/or its affiliates, "The SIP Servlet Tutorial," date last viewed: 03-04-2011. [Online]. Available: http://docs.sun.com/app/docs/doc/820-3007

[40] Oracle Corporation and/or its affiliates, "The SIP Servlet Tutorial," date last viewed: 03-04-2011. [Online]. Available: http://download.oracle.com/docs/cd/E19502-01/821-0203/821-0203.pdf

[41] The Mobicents team, "Mobicents sip servlets," date last viewed: 15-07-2011. [Online]. Available: http://www.mobicents.org/products_sip_servlets.html

[42] VideoLAN, "Video Lan Client," date last viewed: 03-04-2011. [Online]. Available: http://www.videolan.org/vlc/

[43] The MPlayer Project, "MPlayer," date last viewed: 03-04-2011. [Online]. Available: http://www.mplayerhq.hu/

[44] The GStreamer team, "GStreamer," date last viewed: 15-07-2011. [Online]. Available: http://gstreamer.freedesktop.org/

[45] Linphone, "linphone," date last viewed: 15-07-2011. [Online]. Available: http://www.linphone.org/

[46] Larry Doolittle, "ntpclient," date last viewed: 03-04-2011. [Online]. Available: http://doolittle.icarus.com/ntpclient/

[47] The GSL Team, "GSL - GNU Scientific Library," date last viewed: 15-07-2011. [Online]. Available: http://www.gnu.org/software/gsl/

[48] The GSL Team, "GNU Scientific Library – Reference Manual," date last viewed: 15-07-2011. [Online]. Available: http://www.gnu.org/s/gsl/manual/html_node/

[49] Free Software Foundation, Inc, "The GNU C Library," date last viewed: 03-04-2011. [Online]. Available: http://www.gnu.org/s/libc/manual/

[50] Pradeep Padala, "NCURSES Programming HOWTO," date last viewed: 03-04-2011. [Online]. Available: http://www.ibiblio.org/pub/Linux/docs/HOWTO/other-formats/pdf/NCURSES-Programming-HOWTO.pdf

[51] Alexandru Csete, "More GStreamer Tips: Picture-in-Picture Compositing," date last viewed: 03-04-2011. [Online]. Available: http://www.oz9aec.net/index.php/gstreamer/347-more-gstreamer-tips-picture-in-picture-compositing

[52] Wireshark Foundation, "Wireshark · Documentation," date last viewed: 01-08-2011. [Online]. Available: http://www.wireshark.org/docs/

[53] Torsten Löbner, "Dissector for XR IDMS from ETSI TS 182 063 v3.5.2 Annex W," date last viewed: 12-08-2011. [Online]. Available: https://bugs.wireshark.org/bugzilla/show_bug.cgi?id=6163

# Bibliography

[54] R.N. Mekuria, H.M. Stokking, and Dr. M.O. van Deventer, "Automatic measurement of play-out differences for social tv, interactive tv, gaming and inter-destination synchronization," 2011.

[55] Thomas Williams and Colin Kelley, "gnuplot homepage," date last viewed: 17-06-2011. [Online]. Available: http://www.gnuplot.info/

# About the author

Torsten Löbner was born in 1983 in Gera. After his occupational training as Kommunikationselektroniker, he completed a Bachelor's degree in Nachrichten-technik at the Hochschule für Telekommunikation in Leipzig. In addition, the author completed a Master's degree in Informations- und Kommunikationstechnologie at the Hochschule für Telekommunikation in Leipzig. Following the Bachelor's degree, Torsten Löbner gathered experiences in the field of NGN and IMS. During the Master's degree, he specialized in IMS-based IPTV. The Master's thesis 'Implementing ETSI standardised RTCP-based Interdestination Media Synchroni-zation' formed the conclusion of the study and content of this book. The starting point of this work was a project for the analysis and presentation of the current structure of the standardized IMS-based IPTV platform.

www.ingramcontent.com/pod-product-compliance
Lightning Source LLC
LaVergne TN
LVHW080117070326
832902LV00015B/2641